Collagen-Based Biotextiles for Tissue Engineering

Fan Zhang • Daxian Zha • Yihan Huang
Monica Vijay Deshpande • Ting He
Jiyang Chen • Tushar Bambharoliya
Martin W. King

Collagen-Based Biotextiles for Tissue Engineering

Springer

Fan Zhang
Wilson College of Textiles
North Carolina State University
Raleigh, NC, USA

Daxian Zha
Wilson College of Textiles
North Carolina State University
Raleigh, NC, USA

Yihan Huang
Wilson College of Textiles
North Carolina State University
Raleigh, NC, USA

Monica Vijay Deshpande
Wilson College of Textiles
North Carolina State University
Raleigh, NC, USA

Ting He
Wilson College of Textiles
North Carolina State University
Raleigh, NC, USA

Jiyang Chen
Wilson College of Textiles
North Carolina State University
Raleigh, NC, USA

Tushar Bambharoliya
Wilson College of Textiles
North Carolina State University
Raleigh, NC, USA

Martin W. King
Wilson College of Textiles
North Carolina State University
Raleigh, NC, USA

ISBN 978-3-031-71562-4 ISBN 978-3-031-71563-1 (eBook)
https://doi.org/10.1007/978-3-031-71563-1

This Springer imprint is published by the registered company Springer Nature Switzerland AG
The registered company address is: Gewerbestrasse 11, 6330 Cham, Switzerland

About the Book

Tissue engineering has advanced significantly since it was first identified in the 1980s (Niklason & Lawson, 2020). Collagen, the predominant structural protein in the human extracellular matrix (ECM), has been broadly applied in various tissue engineering applications. Collagen in nature, which is the major component in extracellular matrix (ECM), provides mechanical support and protection against damage and injury. It is composed of three polypeptide chains, namely glycine, proline, and hydroxyproline, each of which is composed of at least one Gly–X–Y sequence (where X and Y are either proline or hydroxyproline) structured in a left-handed α-helix. Three of these polypeptide chains are organized into a right-hand triple helix, which is the characteristic structure of collagen. Currently, 28 different types of collagens have been identified, with type I and type II collagen being the most widely used in biomaterials applications. In this review, we introduce collagen as one of the most important biomaterials for tissue engineering applications. We primarily focus on collagen in the form of fibers, textile fabrics, and textile-reinforced hydrogel composites for various tissue engineering applications. In Chap. 1, we introduce the composition, structure, and physical and biochemical properties of collagen. In Chap. 2, different sources, extraction methods, and decellularization methods are discussed. In Chap. 3, we review state-of-the-art fabrication methods of fibrous-collagen-based tissue engineering scaffolds. And then in Chap. 4, we summarize the design of collagen-based textile scaffolds for specific tissue applications that fulfill the requirements of both mechanical performance and biological properties. Overall, we trust this monograph will provide a comprehensive review of the use of collagen-based biotextiles in a range of different tissue engineering applications.

Contents

About the Authors

Fan Zhang, Ph.D., is a Postdoctoral Scholar in the Department of Bioengineering at the University of Washington. Fan received her Ph.D. degree in Fiber and Polymer Science under the direction of Dr. Martin W. King at North Carolina State University, focused on designing and engineering biomaterials for tissue-engineered vascular graft. She has designed a series of novel composite biomaterials that leverage the mechanical strength of fibers and bioactivities of hydrogel and biologics such as extracellular matrix proteins, glycosaminoglycans, and cytokines. Upon receiving her degree, she joined the University of Washington to continue her research in biomaterials and regenerative medicine. Her work bridges several facets of regenerative medicine, including biomaterial engineering, scaffold design, tissue engineering, regenerative immunology, and vascularization. Fan's current research focuses on uncovering the mechanism underlying biomaterial-induced vascular network formation and designing biomaterials to promote vascularization in engineered tissue.

Daxian Zha, Ph.D., received a Ph.D. degree in Fiber and Polymer Science from North Carolina State University. Her research focuses on designing scaffold material for wound healing and tissue engineering. She has been involved in numerous projects, ranging from developing advanced woven and knitted scaffolds for skin tissue regeneration to exploring the use of human-induced pluripotent stem cells for ocular surface reconstruction. With expertise in various biotechniques and a deep understanding of biomedical engineering, she is dedicated to advancing the field and improving patient outcomes through innovative tissue-engineered scaffolds.

Yihan Huang, Ph.D., is a Research and Development Engineer at Cook Medical. She received a B.S. degree in Functional Materials from Donghua University and an M.S. degree in Textile Engineering and a Ph.D. degree in Fiber and Polymer Science (minor in Biomedical Engineering) from North Carolina State University. Her previous work was focused on exploring the potential of a braided scaffold seeded with tendon stem/progenitor cells for rotator cuff tissue engineering, and now is working on identifying clinical needs, conceiving new medical products and solutions, and contributing to the evaluation of new technologies in Vascular R&D.

Monica Vijay Deshpande, Ph.D., is a Product Development Engineer at a biomaterial company in Telford, PA, and holds a Ph.D. in Fiber and Polymer Science from NC State University in Raleigh, NC, USA. Her research focuses on biomaterial development.

Ting He, Ph.D., obtained her M.S. degree in Textile Engineering and Ph.D. in Fiber and Polymer Science from NC State University in Raleigh, NC, USA. Dr. He's industrial experience includes Senior Principal Engineer R&D in Polymers and Textiles Center of Excellence, Edwards Lifesciences, LLC, and End Use Research Representative in The LYCRA Company. Her hands-on experience involves engineering textiles from yarn development to fabric design and their applications in implantable devices or tissue regeneration. Her broad research experiences include 3D cloth component and sutures for heart valve devices, warp-knitted multiple-phase fabric to regenerate muscle-tendon-junction tissue, electro-spun PLGA scaffold as tissue engineering skin scaffold, extrusion of grooved PLA filament to improve cellular response to material surface, and degradation process of textile scaffold in production. Her broad interests also address the eco-system for a textile tissue engineering scaffold to succeed in vitro, including 3D bioreactor design, mechanical stimulation and analysis during tissue regeneration, and non-destructive optical analysis to monitor gene expression and cell penetration inside textile scaffold.

Jiyang Chen, Ph.D., is a Project Manager at a pharmaceutical company in Shanghai, China. He earned his Ph.D. degree from North Carolina State University in Fiber and Polymer Science and received his post-doctoral training at the University of North Carolina Chapel Hill School of Chemistry. His major research focuses on the biodegradable implant and primary cell line establishment.

Tushar Bambharoliya, Ph.D., is a Manufacturing Manager at Becton, Dickinson and Company in Lexington, Massachusetts, USA. Tushar earned his Ph.D. in Fiber and Polymer Science with a minor in Biomedical Engineering under the guidance of Dr. Martin King at North Carolina State University. His research involved designing a mesh-covered stent functionalized with stem cell derived regenerative factors, which aims at enhancing the integration of medical devices within the human body and reduce the risk of adverse reactions and complications, showcasing a remarkable blend of biomedical engineering and textile and polymer technology. Currently, Tushar's primary focus lies in the development and production of soft tissue repair medical devices, available in a variety of shapes and sizes, all incorporating the advanced poly(4-hydroxybutyrate) polymer technology.

Martin W. King, Ph.D., is a Professor of Biotextiles and Textile Technology at North Carolina State University having earned his bachelor's degree in Polymer Technology from Manchester University, UK, and his doctoral degree in Biomedical Engineering from l'Université de Technologie de Compiègne, France. His collaborative work with surgeons and industrial partners throughout North and South America, Europe, and Asia has enabled his research focus to understand the design, evaluation, and clinical performance of a wide range of different implantable medical textile devices, such as sutures, vascular prostheses, hernia and prolapse repair meshes, heart valves, artificial tendons and ligaments, tissue engineering scaffolds, drug delivery membranes, and wound dressings.

Chapter 1
Collagen as an Essential Protein for Tissue Engineering

Collagen is the predominant protein in the extracellular matrix (ECM) and contributes around 30% of the total protein in the human body (Zhang et al., 2015). Due to its inherent presence in ECM, collagen is used extensively in regenerative medicine to mimic the composition and structure of native ECM. In general, collagen presents an arginine-glycine-aspartic acid (RGD) motif, which signals cell adhesion and spreading, and the presence of collagen has been shown to improve cell survival (Gonzalez-Pujana et al., 2020). The fibrillar morphology, viscoelasticity, and rapid stress relaxation of collagen also affect the in vivo cellular response. In this chapter, different aspects of collagen and its roles in tissue engineering will be discussed, including its composition and structure, morphology, mechanical properties, and the conditions that control its degradation. Since the role of collagen is highly tissue dependent, its use will also be discussed in Chap. 4 in tissue-specific applications.

1.1 Composition and Structure of Collagen

The molecular structure of collagen involves a right-hand triple helix composed of three tightly coiled left-hand polyproline II helical polypeptide chains, as shown in Fig. 1.1 (Shoulders & Raines, 2009). The primary sequence of the polypeptide chains consists of repeating Xaa-Yaa-Gly sequences, where Xaa and Yaa can be substituted by any amino acid in the various different types of collagen with the most frequent triplet being Pro-Hyp-Gly (Shoulders & Raines, 2009). Every third amino acid is glycine, which is a small amino acid that fits perfectly inside the helix. The remaining positions in the chain are filled randomly by proline and hydroxyproline (Shoulders & Raines, 2009). The polymer chains are wrapped to form a left-hand polyproline II helix. Three helices twist tightly around one another in a rope-like structure via interchain hydrogen bonding to form a densely packed triple helix. The stabilized interchain hydrogen bonding prevents the collagen molecule

F. Zhang et al., *Collagen-Based Biotextiles for Tissue Engineering*,
https://doi.org/10.1007/978-3-031-71563-1_1

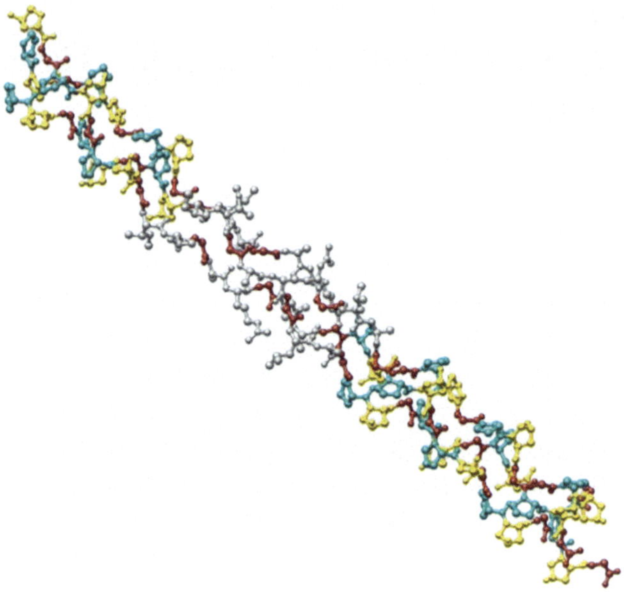

Fig. 1.1 The triple-helix structure of collagen protein. (Reprinted with permission from Avila Rodríguez et al., 2018. © 2017 Wiley Periodicals, Inc.)

from being attacked by other molecules, such as water and collagenases. Normally, the overall molecular size is around 300 nm in length and has a 1.5–2.0 nm diameter (Bella et al., 1994).

There are around 29 different types of collagen. They have been classified into different families. The most frequently used collagen in regenerative medicine is fibrillar collagen and network-forming collagen (León-López et al., 2019). Other families include the fibril-associated collagens with interrupted triple helices (FACITs), the membrane-associated collagens with interrupted triple helices (MACITs), and the multiple triple-helix domains and interruptions (MULTIPLEXINs) (Shoulders & Raines, 2009). Different types of collagen are distributed in different tissues and play distinctive roles in the native ECM as listed in Table 1.1.

The most commonly used types of collagen in tissue engineering applications are the type I to type V collagens. These five types of collagen are discussed below in terms of their distribution in tissue and their structure and function in the development and maintenance of different types of tissues in the human body.

1.1.1 Type I Collagen

Type I collagen is distributed widely in tissues such as the skin, cornea, blood vessels, muscles, tendons, ligaments, cartilage, and bone and is the particular type that is able to form collagen fibers. It is also an essential component in the extracellular

Table 1.1 Various types of collagens[a]

Type	Class	Composition	Distribution
I	Fibrillar	$\alpha1[I]_2\alpha2[I]$	Predominant: dermis, bone, tendon, ligament, meniscus
II	Fibrillar	$\alpha1[II]_3$	Cartilage, vitreous body
III	Fibrillar	$\alpha1[III]_3$	Skin, blood vessels, intestine, cartilage
IV	Network	$\alpha1[IV]_2\alpha2[IV]$	Basement membranes, liver sinusoid
		$\alpha3[IV]\alpha4[IV]\alpha5[IV]$	
		$\alpha5[IV]_2\alpha6[IV]$	
V	Fibrillar	$\alpha1[V]_3$	Widespread: bone, dermis, cornea, placenta
		$\alpha1[V]_2\alpha2[V]$	
		$\alpha1[V]\alpha2[V]\alpha3[V]$	
VI	Network	$\alpha1[VI]\alpha2[VI]\alpha3[VI]$[d]	Widespread: bone, cartilage, cornea, dermis
		$\alpha1[VI]\alpha2[VI]\alpha4[VI]$	
VII	Anchoring fibrils	$\alpha1[VII]_2\alpha2[VII]$	Dermis, bladder
VIII	Network	$\alpha1[VIII]_3$	Widespread: dermis, brain, heart, kidney
		$\alpha2[VIII]_3$	
		$\alpha1[VIII]_2\alpha2[VIII]$	
IX	FACIT[e]	$\alpha1[IX]\alpha2[IX]\alpha3[IX]$	Cartilage, cornea, vitreous
X	Network	$\alpha1[X]_3$	Cartilage
XI	Fibrillar	$\alpha1[XI]\alpha2[XI]\alpha3[XI]$	Cartilage, intervertebral disc
XII	FACIT	$\alpha1[XII]_3$	Dermis, tendon
XIII	MACIT	–	Endothelial cells, dermis, eye, heart
XIV	FACIT	$\alpha1[XIV]_3$	Widespread: bone, dermis, cartilage
XV	MULTIPLEXIN	–	Capillaries, testis, kidney, heart
XVI	FACIT	–	Dermis, kidney
XVII	MACIT	$\alpha1[XVII]_3$	Hemidesmosomes in epithelia
XVIII	MULTIPLEXIN	–	Basement membrane, liver
XIX	FACIT	–	Basement membrane
XX	FACIT	–	Cornea (chick)
XXI	FACIT	–	Stomach, kidney
XXII	FACIT	–	Tissue junctions
XXIII	MACIT	–	Heart, retina
XXIV	Fibrillar	–	Bone, cornea
XXV	MACIT	–	Brain, heart, testis
XXVI	FACIT	–	Testis, ovary
XXVII	Fibrillar	–	Cartilage
XXVIII[f]	–	–	Dermis, sciatic nerve

[a]Information is updated from Shoulders et al. (2009), Bächinger et al. (2010) and Wang et al. (2020)

matrix (ECM). For instance, collagens I and III are the predominant collagens present in the healthy heart, which consists of around 90%–95% of all cardiac collagens (Yokota et al., 2020). Type I collagen is a most suitable biomaterial for fabricating tissue engineering scaffolds or drug delivery vehicles, because of its low risk of inducing an immune response after implantation (Glowacki & Mizuno, 2008).

The typical tropocollagen molecule of type I collagen consists of two alpha1 chains and one alpha2 chain with a continuous triple-helix structure (Fig. 1.2a). These chains combine to form more than 1000 triplet repeat amino acid sequences which are aligned and self-assembled into bundles (Fig. 1.2b) and collagen fibrils (Fig. 1.2c). A collagen fiber, with a diameter ranging from 0.5 to 3 μm, is generated when these fibrils aggregate and become cross-linked (Fig. 1.2d) (Parenteau-Bareil et al., 2010).

1.1.2 Type II Collagen

Type II collagen is made up of three pro-alpha1 chains twisted together to form a triple-stranded structure. But type II collagen is different from type I and III collagen, because of its limited distribution. It is found primarily in particular cartilage and hyaline cartilage (Line et al., 1993; Shoulders & Raines, 2009) and also exists in the nucleus pulposus of the intervertebral disc and in the retina, sclera, and lens of the eye (Line et al., 1993). If one of the genes, COL2A1, which encodes type II collagen mutates due to the heterozygosity, then type II collagen defects will cause various types of mild to severe joint disease, including achondrogenesis II, hypochondrochogenesis, congenital spondyloepiphyseal dysplasia, Strudwick type, and inflammatory and non-inflammatory familial arthropathy (Rimoin et al., 2013).

1.1.3 Type III Collagen

Type III collagen is the second most abundant collagen in soft tissues. It belongs to the fibrillar collagen superfamily and is widely found in connective tissue such as the skin, lung, liver, intestine, and vascular system together with type I collagen. It is worth noting that type I and type III are not found together in the bone (Nielsen

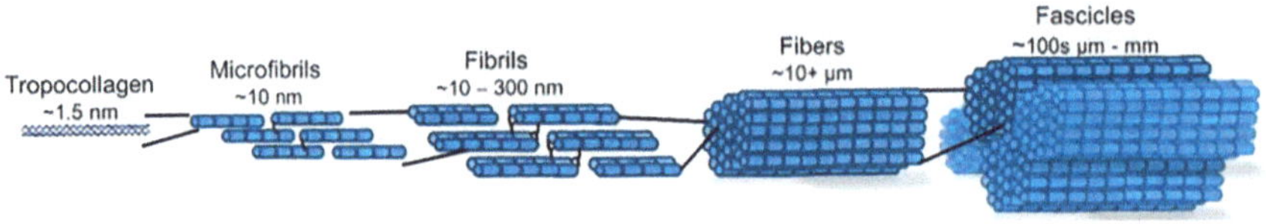

Fig. 1.2 The hierarchical structure of collagen fibers. (Reprinted from Puetzer et al., 2021 under Creative Commons CC-BY license)

& Karsdal, 2016). Type III collagen is composed of three alpha1 chains twisted around each other in a right-hand triple helix (Nielsen & Karsdal, 2016). These chains are built from approximately 1000 residues having the Gly-Xaa-Yaa repeat, which was discussed previously (Shoulders & Raines, 2009).

Type III collagen has strong association with type I collagen as these two types of collagen are major components in the extracellular matrix (ECM) (Nielsen & Karsdal, 2016). Due to the unique covalent cross-links at both the N- and C-terminals adjacent to the triple helical domain, type III collagen serves as a suitable source for undertaking refolding experiments (Bächinger et al., 2010). Since it can be secreted by fibroblasts and mesenchymal cell types, various inflammation-associated pathologies, such as vascular disorders, kidney fibrosis, and hernias, are associated with type III collagen (Nielsen & Karsdal, 2016).

1.1.4 Type IV Collagen

Type IV collagen is primarily found in the basement membrane which has been defined as a complex, specialized extracellular matrix playing an important role in the development and maintenance of many tissues in the body (Feitosa e al., 2011; Sand et al., 2019). Specifically, this network-forming collagen has many binding sites and therefore forms the backbone of the basement membrane which is found between connective tissues and epithelial cells, endothelial cells, and Schwann cells (Bächinger et al., 2010). Type IV collagen functions as a barrier between different types of cells and serves as a cellular support (Sand et al., 2019). As listed in Table 1.1, the chain composition of type IV collagen has six alpha chains, alpha1 to alpha6, which are classified into three subtypes by their gene structure. The first subtype includes alpha1 and two alpha2 chains. Then there is alpha3, alpha4, and alpha5 in a second set and alpha5 and two alpha6 chains in the last set (Shoulders et al., 2009). The subtype with chain composition alpha1 and two alpha2 chains exists in almost all basement membranes. In fact, the basement membrane of kidney glomerulus, the alveoli of the lung, and the neuromuscular synapse all contain the second set with alpha3, alpha4, and alpha5 chains. And the alpha5 plus two alpha6 chains has been found in the basement membranes of the epidermis, smooth muscle cells, mammary glands, and epithelium of the alimentary tract (Bächinger et al., 2010).

1.1.5 Type V Collagen

Type V collagen is a fibrillar collagen that is distributed throughout the body. It has a major effect on tissue quality due to the fact that it is essential for providing optimal fibrillation of type I and III collagens. It has been found that type V collagen is a component of bone matrix, corneal stroma, lung tissue, and the interstitial matrix of muscles. It has been observed as a minor component in the regeneration of scar tissue because it has a major influence on the size and mechanical properties of scars after an acute cardiac arrest (Yokota et al., 2020). The absence of collagen V increases scar size after acute cardiac ischemia. Lack of collagen V alters the mechanical properties of the scar and increases fibroblast activation in an integrin-dependent manner which encourages ECM deposition and leads to the enlargement of the scar (Yokota et al., 2020).

1.2 Topographical Features of Collagen Signaling Pathway and the Effect on Cell Behavior and Tissue Regeneration

Fibrillar collagen contributes to the anisotropy of tissues such as the tendon, bone, myocardium, and blood vessel walls because the morphology affects the cellular response. For instance, it was found that because of its elongated and aligned morphology, endothelial cells (ECs) were able to migrate quickly along the axial direction of fibers (Lai et al., 2012). Aligned collagen nanofibers improve EC outgrowth into a surrounding hydrogel mediated by collagen-binding integrin alpha1 chains, compared to nonaligned collagen scaffolds (Nakayama et al., 2015). Aligned collagen substrates also improve endothelial cell survival in both healthy and ischemic murine models (Huang et al., 2013).

Fiber size also affects EC alignment. Fiber diameters of about 30 nm compared to 100 nm have a more significant effect in guiding cell alignment and regulating the EC inflammatory response (Huang et al., 2013). It has also been shown that fiber substrate morphology helps to maintain the quiescent phenotype of vascular smooth muscle cells (VSMCs) (Koyama et al., 1996).

1.3 Mechanical Properties of Collagen Mediate Specific Cellular Responses

Tissues at different locations assume different mechanical properties and thus provide unique mechanical signals to the resident cells. ECM presents different stiffness values in different tissues, ranging from 1kPa in the brain to about 10 kPa in the human heart and about 30 kPa in the pre-calcified bone under physiological conditions (Birukova et al., 2013; Lee et al., 2019). These differences in mechanical properties provide instructive signals to program cell adhesion, migration, and proliferation in different tissues. The stiffer substrates promote the production of focal adhesion related proteins and enhance the assembly of mesenchymal stem cells (MSCs) compared to softer substrates as shown in Fig. 1.3 (Engler et al., 2008). Stiffer substrates also encourage MSC elongation (Engler et al., 2008). Pathological fibrotic tissue is typically harder and stiffer than healthy tissue, which

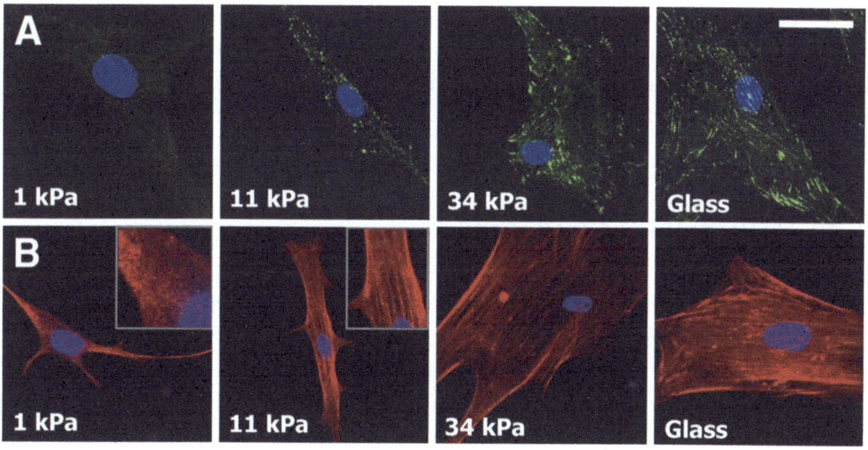

Fig. 1.3 Collagen stiffness alters cell adhesion and spreading. Scale bar = 20 μm. (**a**) Paxillin-labeled adhesions and (**b**) F-actin organization of human MSCs were both modified on stiffer substrates compared to softer substrates. (Reprinted with permission from Engler et al., 2006. Copyright © 2006 Elsevier, Inc.)

can result in the malfunction of attached and embedded cells. For instance, in an atherosclerotic artery in a rabbit, the stiffness of the matrix can be as high as 100 kPa (Birukova et al., 2013). In fibrotic cardiac tissue after a myocardial infarction, myocardiocytes lose their ability to beat rhythmically on the stiffened matrix (Engler et al., 2008).

External mechanical forces are transferred by integrin adhesion and modification to the actin cytoskeleton arrangement, which activates Yes-associated protein (YAP) transcriptional coactivator with a PDZ-binding domain (TAZ) (Lee et al., 2019). YAP was found to enhance cell proliferation and promote cardiac cell contractility but was also observed to decrease their maturation (Lee et al., 2019). Chaudhuri et al. demonstrated that the viscoelastic properties of ECM proteins regulate cell spreading by affecting actomyosin contractions and clustering of adhesion ligands. Faster stress relaxation on ECM proteins facilitates faster modifications in cell shape and encourages cell spreading and proliferation (Chaudhuri et al., 2016).

In addition to cell adhesion and proliferation, matrix mechanical properties, such as stress relaxation, elasticity, and stiffness, also instruct cells to develop different phenotypes. Soft matrices that mimic the brain ECM encourage neurogenic differentiation of mesenchymal stem cells (MSCs), whereas stiffer matrices that mimic aligned muscle tissue promote a myogenic phenotype (Engler et al., 2006). The intermediately rigid matrices that mimic bone ECM are known to support osteogenic growth (Engler et al., 2006), and rapid stress relaxation has also been reported to inhibit adipogenesis of MSCs but at the same time induce osteogenesis (Chaudhuri et al., 2016). Trappmann et al. found that on the softer collagen coated polyacrylamide hydrogel substrates, the migration of keratinocytes was reduced compared to the stiffer bare polyacrylamide surfaces, while differentiation was promoted

(Trappmann et al., 2012). Integrin clustering was less evident on the soft substrates, which reduced activation of the extracellular signal-related kinase (ERK)/mitogen-activated protein kinase (MAPK) signaling pathway, resulting in decreased differentiation (Trappmann et al., 2012).

1.4 Degradation of Collagen

The compact three-dimensional structure of collagen results in a slow degradation process that is not readily initiated. Only specific enzymes, known as collagenases, can mediate the hydrolysis of collagen, and the majority of these enzymes are zinc dependent matrix metalloproteinases (MMPs) (Motta et al., 2012).

MMPs are a large family of calcium-dependent zinc-containing endopeptidases, which are responsible for the degradation of the ECM and tissue remodeling, including collagen, elastin, gelatin, matrix glycoproteins, and proteoglycan. There are at least 26 human MMPs regulated by hormones, growth factors, and cytokines (Verma & Hansch, 2007). They are classified as collagenases, gelatinases, stromelysins, and matrilysins, and among these, there are four collagenases, namely, MMP-1, MMP-8, MMP-13, and MMP-18, otherwise known as collagenase-1, neutrophil collagenase, collagenase-3, and collagenase-4, respectively (Verma & Hansch, 2007). These MMPs are mosaic proteins, each composed of modular inserts and domains. From N- to C-terminus, it is composed of a signal peptide for secretion, about 80-residue zymogenic pro-peptide, about 165-residue zinc- and calcium-dependent catalytic domain (CAT), 15- to 65-residue linker, and about 200-residue hemopexin-like domain (HPX) for collagen binding, as well as pro-MMP activation and dimerization region (Fig. 1.4) (Tallant et al., 2010). The pro-peptide domain prevents the MMP from undertaking extensive noncontrolled protein degradation, while the catalytic domain contains two zinc ions, a structural one, and the other one is in the active site, essentially to promote the proteolytic activity of MMPs (Motta et al., 2012).

The mechanism of proteolysis by MMPs is shown in Fig. 1.5. The scissile amide carbonyl link is attracted to the active site of the Zn^{2+} ion, and the carbonyl group is attacked by a water molecule which is hydrogen bonded to a glutamic acid residue and also attracted to the Zn^{2+} ion. The water molecule donates a proton to the Glu residue, followed by the Glu residue transferring a second proton from the water

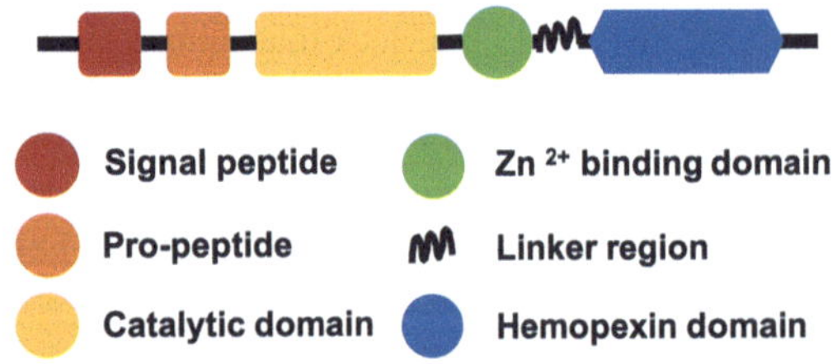

Fig. 1.4 Schematic representation of the organization of collagenase MMP domains. (Motta et al., 2012)

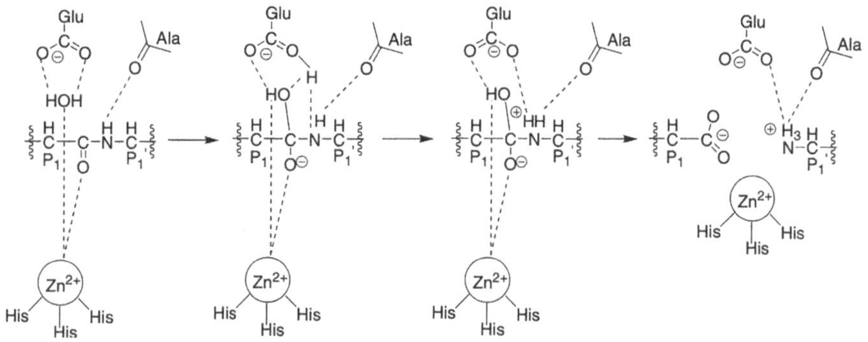

Fig. 1.5 Reaction mechanism for the proteolysis by MMPs. (Reprinted with permission from Lovejoy et al. 1994. Copyright © 1994, American Chemical Society)

molecule to the nitrogen of the scissile amide, which results in peptide bond cleavage. During the proteolysis, the Zi^{2+} plays a role as stabilizer with the presence of a negative charge on the carbon of the scissile amide (Verma & Hansch, 2007).

In the process of fabricating collagen-based tissue engineering scaffolds, the degradation of collagen results in a gradual modification of its mechanical properties and morphology. However, after implantation, the rate of degradation of the collagen scaffold is faster than for synthetic polymeric materials, which contributes to the rapid resolution of inflammation and reduced calcification around the scaffold (Sugiura et al., 2017).

Chapter 2
Sourcing, Extracting, and Processing Collagen Raw Materials

In this chapter, we describe the different sources of collagen raw materials and discuss the effect of different extraction and processing methods on the quality and properties of the resulting collagen material.

2.1 Collagen Raw Material Sourcing for Medical Uses

Various animal tissues can be used for the extraction of collagen. For example, animal sources such as bovine, porcine, and marine organisms, like scarf fish, have the advantage of being low cost and easy to acquire (Avila Rodríguez et al., 2018).

Bovine skin, bone, tendons, and ligaments are the primary industrial sources of collagen. Specifically, type I collagen is extracted from cow skin and tendons, and type II collagen is obtained from both bovine nasal as well as articular cartilage. In addition, type IV collagen is harvested from the placental villi (Strawich & Nimni, 1971). However, 3% of the human population is known to be allergic to bovine collagen. In addition, there are potential risks of transmissible diseases such as bovine spongiform encephalopathy (BSE), transmissible spongiform encephalopathy (TSE), and foot-and-mouth disease (FMD) from these sources of collagen. That is why collagen from Australia and New Zealand are preferred since the cattle in those countries were not exposed to foot-and-mouth disease.

Porcine skin and bones are two other major sources of collagen. Since porcine collagen is similar to human collagen, it does not cause as many hypersensitivity reactions as bovine collagen does. However, porcine collagen may include potential contamination due to zoonosis, and in some countries, it is unacceptable for human use on account of religious beliefs (Silvipriya et al., 2015).

Rat tail tendon is also a valuable and popular source of collagen due to its high content of type I collagen, and it has been widely used in the extraction and preparation of commercial collagen products available on the market (Rittié, 2017). Marine

F. Zhang et al., *Collagen-Based Biotextiles for Tissue Engineering*,
https://doi.org/10.1007/978-3-031-71563-1_2

Table 2.1 Collagen sources, sterilization methods, and commercial applications (Meyer, 2019)

Animal species	Tissue	Sterilization method	Product applications
Porcine	Dermis	Gamma	Dental, burns IIb–III, trauma, reconstructive and surgical wounds, soft tissue repair, hernia and abdominal wall repair, hemostasis
	Small intestine submucosa	Ethylene Oxide (EtO)	Soft tissue augmentation, burns IIb–III, trauma, reconstructive and surgical wounds
	Pericardium	EtO	Soft tissue reconstruction
	Gelatin	Gamma	Hemostasis
Bovine	Tendon	EtO	Dental
	Pericardium	Aseptic production	Spinal and cranial repair
	Gelatin	Gamma	Hemostasis
	Dermis	Gamma, filtration	Hemostasis, burns IIb–III, trauma, reconstructive and surgical wounds, soft tissue augmentation in plastic surgery
Equine	Tendon	EtO	Dental, hemostasis
Human	Dermis	Aseptic production	Dermis replacement, soft tissue reconstruction

life is another option for a feasible source of collagen raw material, as it can be extracted from sponges, fish, and jellyfish. These sources of collagen are also more widely used in industry (Parenteau-Bareil et al., 2010).

In the following Table 2.1, various animal species that serve as a source for collagen are listed together with the commercial applications and the method used for sterilization (Meyer, 2019).

2.2 Extraction

In general, type I collagen is extracted via salt precipitation, acid-based digestion, and enzymatic isolation using neutral saline solution, an acidic solution such as acetic acid or hydrochloric acid, as well as an acid-enzyme mixture such as pepsin in acetic acid (Chowdhury et al., 2018). Here, we introduce three methods of extracting collagen from different sources.

2.2.1 *Extraction from Porcine Tendon*

Porcine tendon is washed and incubated in an acid solution to swell the tissue. Then, the tendon is ground into small particles for easier penetration of the acid. This facilitates dissolution of the collagen and removal of the non-dissolvable contaminants such as fat and cells. The dissolved proteins are precipitated and collected so as to obtain purified collagen. A viral inactivation step, shown to inactivate viruses and bacteria, is then performed at the end of the purification process (Matinong et al., 2022; Wang et al., 2021).

2.2.2 Extraction from Rat Tail Tendon

Because rat tails are readily available in experimental quantities, a common source of type I collagen can be obtained from rat tail tendons. The advantage of this approach is that the concentration of type I collagen is higher than in other tissues, and the protein is easy to isolate with less contamination (Rittié, 2017). Unlike the skin, rat tail tendons contain less blood, lipids, and other non-collagen proteins which would require more steps in the purification process. The process starts with removing the rat tail skin and dissecting the tendon. To eliminate contaminants, such as fat and cellular components, the tendon fiber bundles are dissolved in a weak acid solution and centrifuged. After removing the non-dissolvable contaminants, type I collagen is precipitated by adding salts. The salts and excess acetic acid are removed by dialysis against an acetic acid solution at the desired concentration. The purified collagen solution can either be preserved in solution form or lyophilized for long-term storage. This relates to the concentration, preservation, and further sterilization for cell culture applications (Rittié, 2017). The end product should be validated by biochemical analysis to demonstrate successful extraction. The recommended analysis would be sodium dodecyl sulfate-polyacrylamide gel electrophoresis (SDS-PAGE) due to the unique pattern of migration of type I collagen. Normally, typical migration bands of type I collagen can be seen by running it in an 8% Coomassie brilliant blue-stained SDS-PAGE under reducing conditions. The results from a previous study showed two monomeric alpha-chains (alpha1 and alpha2), dimeric beta-chains, and one trimeric gamma-chain. Based on the analysis, the image showed that the alpha1 and alpha2 chains had average molecular weights of approximately 130 and 140 kDa, respectively. However, there were extra bands observed between the beta-chains and gamma-chains compared to the results from commercial type I collagen. This is most likely due to other protein components besides type I collagen (Xie et al., 2021).

2.2.3 Extraction from the Human Placenta

Collagen can also be extracted from the human placenta using acetic acid and sodium chloride (NaCl). First, samples are washed with distilled water to remove any residual blood, and then they are chopped into small 1–2 cm pieces. The pieces are then mixed with 0.1 N sodium hydroxide (NaOH) at a sample/alkali solution w/v ratio of 1:10 to remove any non-collagen proteins. The mixed solution is stirred

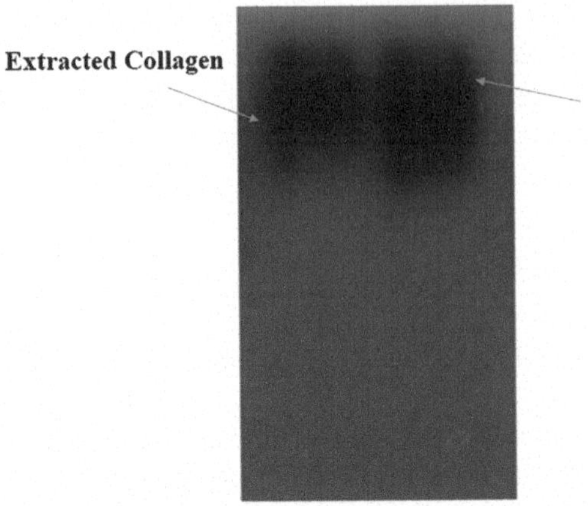

Fig. 2.1 Western blot analysis using anti-His-Tag antibodies for identifying type I collagen expression. (Adapted from Karami et al., 2019 under Creative Commons Attribution License)

for 6 h and the alkali solution is renewed every 2 h. The collagen sample can then be stored in a −20 °C freezer until ready for use. The frozen sample is thawed by washing in cold distilled water (4 °C) for 2 h. To remove any fat, the sample is immersed overnight in 10% butyl alcohol with a w/v ratio of 1:8, rinsed in distilled water, and then exposed again overnight to 0.06 M acetic acid using a w/v of 1:25. After that, the solution is filtered through cheese cloth four times. In order to precipitate the collagen, NaCl is added to the solution and stirred for another overnight at room temperature to achieve a final concentration of 3M at pH 7. The precipitate is then centrifuged to dissolve any remaining sample in 2% acetic acid, and the final step is to dialyze the solution against distilled water for 5 days using a cellulose membrane prior to freeze drying for storage (Karami et al., 2019). After extraction, Western blot analysis can be performed to inspect the difference between the extracted collagen and the standard type I collagen. As shown in Fig. 2.1, this method demonstrates that there is no significant difference between the extracted collagen and the type I collagen standard (Karami et al., 2019).

2.3 Decellularization

During or after the extraction of collagen, it is important to remove any cellular fragments or genetic material that would provoke an immune response once the collagen biomaterial was implanted in a human patient. This process is called decellularization. There are three main strategies for undertaking decellularization, namely, using physical, chemical, or enzymatic methods. A combination of these methods can also be used to optimize the outcome.

Physical decellularization typically relies on hydrostatic pressure and supercritical carbon dioxide (CO_2) at high pressure to remove DNA fragments. However, hydrostatic pressures over 600 MPa have been found to denature ECM proteins (Gilpin & Yang, 2017). An alternative approach using a series of freeze-thaw cycles between −80 and 37 °C can remove cells from tissue samples, but the DNA remains.

Chemical decellularization uses surfactants, acids, and alkalis. Surfactants destroy the phospholipid membrane of cells and separate them from the ECM proteins. Commonly used surfactants, such as Triton X-100 and sodium deoxycholate, can remove cellular components while preserving the matrix (Kasimir et al., 2003). Sodium dodecyl sulfate (SDS) is another effective surfactant, but it has been found to be cytotoxic (Fernández-Pérez & Ahearne, 2019), and therefore it requires extensive washing before further processing (Gilpin & Yang, 2017). SDS is also reported to modify the structure of matrix proteins (Kasimir et al., 2003). Acids and alkalis lyse cells by dissolving the cell membrane. Ethylenediaminetetraacetic acid (EDTA) is used in combination with trypsin to degrade cell-matrix adhesion proteins. The disadvantage of using peracetic acid is due to its corrosive nature which stiffens ECM proteins (Gilpin & Yang, 2017). Ammonium hydroxide can be used, but it leaves a negative charge on the collagen molecule and causes it to swell, which compromises its mechanical properties (Fernández-Pérez & Ahearne, 2019).

Enzymatic methods are less traumatic and more efficient compared to the methods described previously. A commonly used enzyme is trypsin, which cleaves those proteins that anchor cells to the matrix. Typically, it is used together with EDTA as mentioned previously (Fernández-Pérez & Ahearne, 2019). Deoxyribonuclease (DNase) and ribonuclease (RNase) break down DNA and RNA, respectively (Gilbert et al., 2006). However, enzymatic treatment is relatively more expensive.

Each of these strategies has its advantages and limitations, and therefore in order to optimize the decellularization process, it is necessary to apply them in a balanced manner so as to achieve complete removal of cellular and genetic fragments and avoid damage to the collagen structure which will compromise its inherent properties.

Chapter 3
Fabrication and Processing of Collagen-Based Biotextiles for Tissue Engineering Applications

Collagen is an essential biomaterial in many different tissue engineering applications. We have discussed its chemical composition, physical properties, biocompatibility, and biodegradability in previous chapters, as well as alternative sources and extraction methods as a raw material. Despite the superior biological performance of collagen and its derivatives, collagen scaffolds fabricated by traditional techniques lack adequate mechanical performance (Cheng et al., 2008). This is primarily due to the unsuccessful restoration of the helical configuration of the natural collagen macromolecule and the lack of an aligned and fibrous morphology of the collagen bundles in traditionally fabricated collagen gels, sponges, and membranes. This results in their inferior mechanical performance, poor processability, and scalability, which limits their potential for both commercialization and clinical application (Learn et al., 2019).

By applying textile technology to collagen, one can significantly improve its mechanical properties. The inherent requirement for processing collagen on textile machinery is to have available a continuous collagen fiber or yarn. A fiber or yarn is a material with a high length-to-diameter ratio (Mostafizur Rahman et al., 2023), also known as a slenderness ratio. This unique feature gives fibers and yarns comparatively strong tensile properties per unit diameter and at the same time good flexibility to fabricate and process into two- or three-dimensional structure.

Biotextiles, fiber-based biomaterials designed to function in a biological environment, involve a textile structure and fabrication technology that have been widely used in commercial implantable medical devices, such as surgical sutures, hernia meshes, vascular prostheses, and artificial skin grafts. Biotextiles leverage the anisotropy of fibrous materials and the compliance of textile structures and have long provided superior biomaterial candidates for tissue engineering applications.

In addition to providing an anisotropic structure, biotextiles offer excellent mechanical properties compared to traditional fabrication methods and provide morphological signals that resemble the fibrous extracellular matrix (ECM). Its highly porous structure also allows nutrient transmission, cell migration and

F. Zhang et al., *Collagen-Based Biotextiles for Tissue Engineering*,
https://doi.org/10.1007/978-3-031-71563-1_3

proliferation, and ECM deposition, all of which are critical features for tissue regeneration. Natural ECM is a fibrous composite, with some fibers well oriented in a uniaxial alignment for load-bearing purposes and others randomly oriented for different and various tissue functions.

In addition, biotextiles also allow variation in scaffold architecture and structural morphology so as to achieve complex architectural designs and fulfill different requirements for a wide variety of tissue engineering applications. This can be achieved by changing the fiber diameter and its cross-sectional shape, the degree of alignment, and textile fabrication technology, including weaving, knitting, braiding, nonwoven production, and electrospinning.

In this chapter, we provide a comprehensive review of the state-of-the-art fabrication methods and processes for collagen-based biotextiles, including electrospinning, electrochemical alignment, fiber extrusion, surface modification, collagen yarn spinning, and the preparation of fiber-reinforced composites. Some of these methods improve the mechanical performance and durability of resulting collagen biomaterials by enhancing the molecular alignment of collagen macromolecules in one direction, while other methods leverage the mechanical properties of the hybrid materials.

3.1 Electrospinning Collagen Fibers

Electrospinning is a method that creates continuous fibers or filaments with diameters in the range of submicrons to nanometers, using an electric force to attenuate and draw charged threads of a polymer solution or polymer melt (Deitzel et al., 2001). Materials with a nanofibrous structure exhibit several unique characteristics when compared to their microfiber counterparts. For example, they have a large surface area-to-volume ratio. This ratio can be as large as 103 times for a nanofiber compared to that of a microfiber. In addition, it has tunable porosity and flexibility. The advantage of high surface-to-volume ratio and ease of surface functionalization leads to nanofibers having superior mechanical performance in terms of stiffness and tensile strength (Huang et al., 2003). And based on these unique and inherent properties, a collagen nanofiber mesh produced by electrospinning becomes an attractive candidate for a tissue engineering scaffold due to its close resemblance to native tissue architecture.

Compared to other techniques, electrospinning is a relatively simple system. The typical schematic diagram of the basic setup for electrospinning is illustrated in Fig. 3.1. The basic components of electrospinning include a high-voltage power supply, a capillary tube with a pipette or needle of small diameter at the end, and a charged collector. Over time, the structure of the web of collected nanofibers has a similar appearance to a typical nonwoven sample as seen in the scanning electron microscope (SEM) image in Fig. 3.1. During the process, the polymer solution is installed in the container, and a high voltage is applied to create an electrically charged jet of polymer solution as it leaves the pipette or needle. The formation of

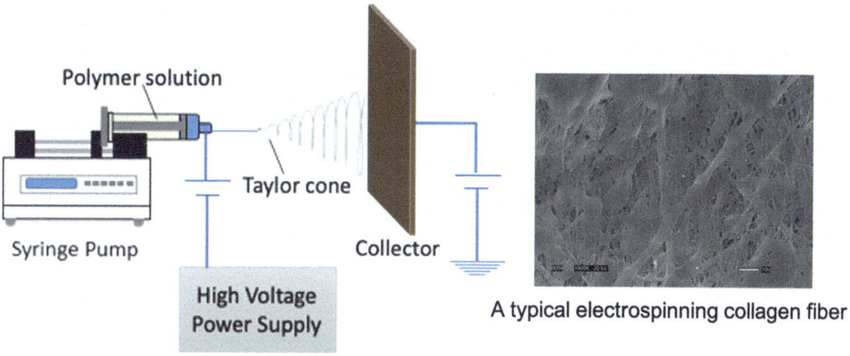

Fig. 3.1 Schematic diagram of a typical electrospinning process and image of a electrospun web of collagen fibers

a Taylor cone at the tip of the needle is subjected to increasing electric potential. So when the surface tension forces of the polymer solution are exceeded by the applied electrical field, a charged jet of polymer solution is ejected from the apex of the Taylor cone. This charged polymer jet experiences a whipping motion and is attenuated continuously due to the electrostatic field prior to being deposited on a grounded collector. During its travel through the atmosphere, the polymer solution solvent evaporates, leaving behind the dry fibers on the collecting device (Liu et al., 2004).

The first-time electrospinning was used to produce a fibrous web as a scaffold for tissue engineering was when Annis et al. electrospun a polyurethane mat as a vascular prosthesis in 1978 (Annis et al., 1978). However, collagen has been widely used as a tissue engineering scaffold due to its structural support to cells and its ability to provide biological and mechanical cues for cell proliferation and tissue formation. The first successful electrospinning of a pure collagen nanofiber web using 1,1,1,3,3,3 hexafluoro-2-propanol (HFIP) as the solvent was reported by Matthews et al. in 2002 (Matthews et al., 2002). Since that time, an increasing number of publications have achieved similar success in producing a collagen nanofibrous matrix by electrospinning a biodegradable and biomimetic scaffold for various tissue engineering applications. There is an additional discussion regarding electrospun collagen scaffolds for tissue repair in Chap. 4.

3.1.1 Effect of Electrospinning Conditions on Collagen Fiber Formation

There are several factors that influence the diameter and morphology of electrospun fibers and yarns, and they include the electrospinning variables, the spinning solution properties, and the environmental parameters. The electrospinning variables include the applied voltage, the distance between the capillary and the collector, the solution flow rate, and the needle diameter. The solution parameters are more focused on the type of solvent, the polymer concentration and viscosity, the surface tension, and the solution conductivity. The electrospinning environmental parameters include the ambient relative humidity and temperature. Table 3.1 presents a general picture of how these variables and parameters directly affect the generation of electrospun fibers and filaments (Law et al., 2017).

Therefore, although the electrospinning system is easy to assemble and operate, in order to optimize the electrospinning conditions and consistently fabricate

Table 3.1 Summary of the effects on electrospinning variables on fiber formation

Parameters		Effects on fiber formation	References
Processing conditions	Distance between capillary tip and collector	Generation of beads with too short or too long distance, optimum distance required for uniform fiber formation	Megelski et al. (2002), Ki et al. (2008) and Homayoni et al. (2009)
	Solution flow rate	Larger fiber diameter with increase in flow rate, generation of beads with too high flow rate	Megelski et al. (2002), Liu et al. (2009) and Zhang et al. (2005)
	Applied Voltage	No effect on fiber diameter	Gu et al. (2005) and Şener et al. (2011)
		Larger fiber diameter with increase in voltage	Liu et al. (2009, 2011)
		Smaller fiber diameter with increase in voltage	Beachley and Wen (2010) and Rodoplu and Mutlu (2012)
Polymer solution parameters	Surface tension	Less beads with decrease in surface tension	Fong et al. (1999) and Yang et al. (2004)
	Conductivity	Smaller fiber diameter and less beads with increase in conductivity	Zheng et al. (2006), Huang et al. (2006) and Lin et al. (2004)
	Polymer concentration & viscosity	Larger fiber diameter and less beads with increase in concentration and viscosity	Veleirinho et al. (2008), Sukigara et al. (2003) and Mo et al. (2004)
	Polymer molecular weight	Larger fiber diameter and less beads with increase in molecular weight	Tao and Shivkumar (2007), Koski et al. (2004) and Kim et al. (2004)
Ambient parameters	Temperature	Smaller fiber diameter with increase in temperature	Zhou et al. (2006), Wang et al. (2007) and Mit-uppatham et al. (2004)
	Relative humidity	High humidity results in pore formation within the fiber	Nezarati et al. (2013), Kim et al. (2005), Medeiros et al. (2008), Medeiros et al. (2008) and Francis et al. (2010)

[a]Reproduced with permission from Law et al. (2017). Copyright © 2017, The Korean Tissue Engineering and Regenerative Medicine Society and Springer Science Business Media B.V

electrospun nanofibers, it is necessary to thoroughly understand the effects of these variables and parameters on the production of a collagen nanofiber web.

3.1.1.1 Effect of Solvents on Electrospun Collagen Nanofibers

The selection of the solvent is an important factor when electrospinning collagen, since a good solvent can facilitate the fabrication of smooth and bead-less nanofibers without degrading the protein. The selection of 1,1,1,3,3,3 hexafluoro-isopropanol (HFIP) solvent is a popular choice for collagen electrospinning due to its inherent physical and chemical properties that provide unique modes of reactivity (Colomer et al., 2017). Collagen can be completely dissolved in HFIP at room temperature. And since the boiling point of HFIP is relatively low, its volatility and high evaporation rate will lead to reduced solvent waste during the attenuation of the nanofibers as they are extruded from the needle tip and travel to the collector (Colomer et al., 2017). Similar to HFIP, 2,2,2-trifluoroethanol (TFE) is another option as a solvent for collagen due to its high volatility (Law et al., 2017). Although HFIP and TFE have advantages in generating collagen fiber formation, they are both toxic and hazardous when inhaled or in contact with the human skin and the environment. It has been reported that HFIP can denature the triple-helical structure of collagen which leads to the formation of gelatin (Chakrapani et al., 2012). Zeugolis et al. found that when collagen is electrospun into nanofibers from fluoroalcohols, such as HFIP or TFE, a 45% loss is observed in the triple-helical structure of collagen as measured by circular dichroism (CD) spectroscopy (Zeugolis et al., 2008b).

Acetic acid was also studied as a potential solvent for electrospinning collagen. However, pure collagen can't be electrospun when dissolved in 100% acetic acid. Collagen nanofibers can only be electrospun when polycaprolactone (PCL) is blended with acetic acid as an environmentally benign solvent (Chakrapani et al., 2012).

A mixture of phosphate buffered saline and ethanol solution has been shown to be another successful solvent for electrospinning pure collagen scaffolds. This benign solvent mixture has improved electrospinnability compared with other water-based polymeric solutions by lowering the surface tension, enhancing the conductivity, and facilitating the formation of a stable Taylor cone. Dong et al. have electrospun type I collagen webs from pH7 buffer/ethanol solutions and obtained a highly oriented fibrous structure that was readily cross-linked by 1-ethyl-3-(3-dimethylaminopropyl) carbodiimide hydrochloride (EDC) (Dong et al., 2009).

3.1.1.2 Effect of Collagen Type and Source on Electrospun Collagen Nanofibers

As mentioned previously, collagen can be isolated from various sources and electrospun into collagen nanofibers which serve as primary structural components of the ECM (Matthews et al., 2002) Since Matthew et al. first demonstrated that 100%

type I and type III collagen could be successfully electrospun using HFIP as the solvent, more and more researchers have explored electrospinning 100% pure type I, type II, type III, and type IV collagen using different concentrations of collagen solution (Matthews et al., 2002). Type I collagen solution with a concentration of 8% by weight has been electrospun into a nonwoven web of randomly arranged continuous collagen fibers or filaments, which have subsequently been cross-linked chemically by glutaraldehyde vapor. An aqueous 0.1M glycine solution was applied later in order to remove any unreacted aldehyde groups. The mechanical properties, cytocompatibility, cell behavior, and interaction with collagen were examined to estimate the potential of this approach to electrospinning collagen. The results showed a 0.2-mm-thick type I collagen nanofibrous matt had sufficient tensile strength in excess of 10 MPa. When exposed to normal human keratinocytes, the collagen nanofiber matts responded with positive functional activity (Rho et al., 2006). Similarly, type II collagen has been successfully electrospun for tissue engineering of articular cartilage and intervertebral disks (Matthews et al., 2016). In a study to develop a lyophilized sternal cartilage in the chest of a chicken, type II collagen was suspended in 1,1,1,3,3,3 hexafluoro-isopropanol (HFIP) at concentrations of 0.04 and 0.1 g/mL, and these samples were electrospun into nonwoven fibrous matts composed of type II collagen fibers with diameters ranging from 110 nm to 1.8 μm, respectively (Matthews et al., 2016). As mentioned previously, type III collagen is composed of three alpha I chains which vary in size from 30 nm to 130 nm in diameter (Matthews et al., 2002). According to Matthews et al., this type III collagen was isolated from human placenta and dissolved in HFIP at various concentrations for electrospinning. They compared the structural properties of the resulting type III collagen with electrospun type I collagen and found significant differences in diameter for these two isotopes. In the same study, by comparing type I calfskin with type I placenta collagen, they were able to evaluate the impact of the electrospinning process on the formation of the fibrous matt. The electrospun collagen fibers isolated from human placenta were less uniform and had a larger range of diameters compared to fibers spun from calf skin; see Fig. 3.2 (Matthews et al., 2002).

Collagen, being an animal derived material, requires reliable and reproducible purification methods and standard analytical assays for characterization. And given that collagen is obtained from a complicated extraction procedure, Zeugolis et al. found the properties of electrospun collagen fibers and their solubility in the same solvent vary from one sample to another even after extraction from the same source. (Zeugolis et al., 2008a). For example, high purity type I bovine dermal atelocollagen solutions obtained from two different companies, Koken (Tokyo, Japan) and Symatese Biomateriaux (Chaponost, France), generated electrospun fibers with mean diameters of 665 ± 150 and 495 ± 78 mm, respectively. The SDS-PAGE analysis was performed to estimate the solubility in HFIP solvent, and the results demonstrated differences in collagen content. For example, the collagen content in the Sigma-Aldrich product (Singapore; calf skin; catalog no. C-9791; lot no. 05K7042) and the Koken sample were relatively high compared to the insoluble or low collagen content observed in the Fluka product (Singapore; bovine Achilles tendon;

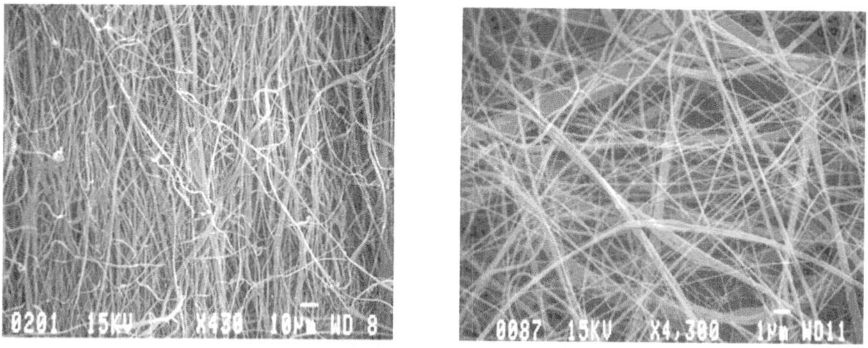

Fig. 3.2 SEM images of calfskin electrospun type I collagen (left) and electrospun type III collagen from human placenta (right). Average fiber diameter for calfskin type I collagen is 100 ± 40 nm compared to 250 ± 150 nm for the electrospun type III collagen fibers. (Adapted with permission from Matthews et al., 2002. Copyright © 2002, American Chemical Society)

catalog no. 27662) and the Yi Erkang sample (Beijing, China; bovine Achilles tendon) (Zeugolis et al., 2008a).

3.2 Electrochemical Alignment

Isoelectric focusing is an electrophoresis technique that separates proteins based on their isoelectric point (pI), where the pl is the pH at which a protein has no net charge and is not moved by an electric field. Type I collagen is an ampholytic protein which exhibits −0.8 coulomb charge at pH 3 and +0.8 coulombs at pH 11. This enables the molecule to migrate at its isoelectric point under a pH gradient induced by an external electric field (Dewle et al., 2020). Different electrochemical reactions occur at the anode and cathode, which generates the pH gradient between the electrodes. As a result, collagen molecules can migrate toward their isoelectric point at pH 8, resulting in the self-assembly of highly aligned collagen bundles near the cathode (Dewle et al., 2020).

Taking advantage of this unique characteristic of collagen, Akkus et al. have developed a customized rotating electrode electrochemical alignment device (REEAD) which spins a continuous and highly aligned collagen monofilament (Akkus et al., 2018) by relying on a stable electrical potential across its rotating electrode (Younesi et al., 2014) (Fig. 3.3).

The collagen is dissolved in an acidic solvent and supplied by a syringe pump to the upper bath, where it gets trapped in the grooves on the surface of the rotating electrodes. Under the influence of the electrical potential, the solution forms a pH gradient between the two electrodes, which aligns the collagen molecules in the axial direction. As the solvent evaporates, the collagen fibers solidify and form continuous filaments (Fig. 3.4a) (Kew et al., 2011).

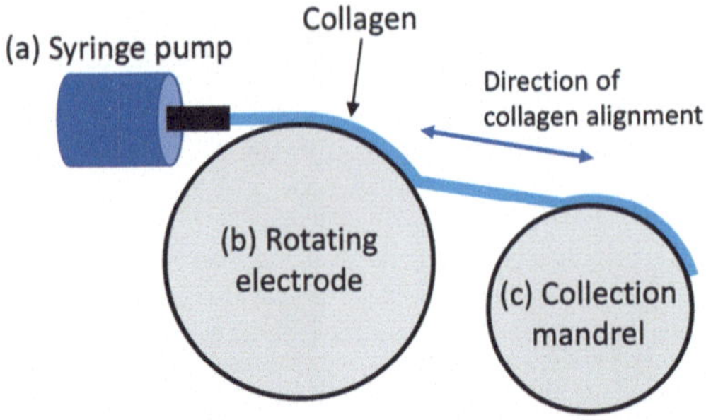

Fig. 3.3 Schematic diagram of the rotating electrode electrochemical alignment device (REEAD) for continuous electro-compaction of a collagen monofilament: (**a**) syringe pump, (**b**) rotating wheel with electrodes on the two edges, and (**c**) rotating collection mandrel. (Reprinted under Creative Commons Liscence from Carr et al., 2022. Copyright© 2022 By Carr et al.)

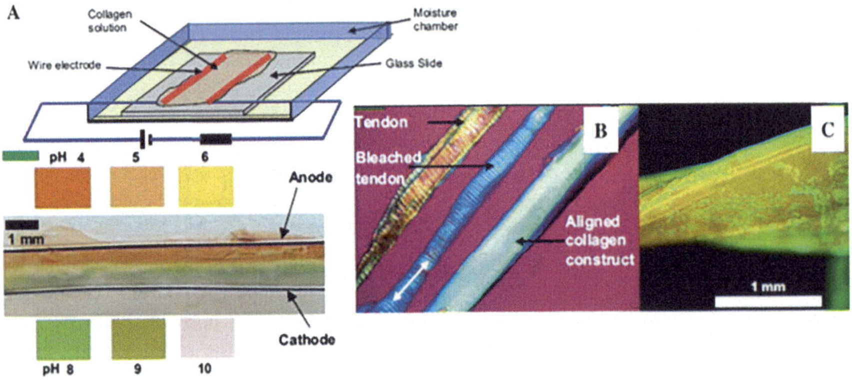

Fig. 3.4 (**a**) Diagram of the electrochemical device that generates a pH gradient and aligns the collagen molecules to produce continuous collagen filaments. (**b**) Under polarized light microscopy, ELAC shows a similar degree of orientation to native tendons. (**c**) Tendon-derived fibroblasts extensively populate ELACs after 7 days of in vitro culture. Reprinted with permission from (Kew et al., 2011). (Reprinted with permission from Kew et al., 2011. Copyright © 2011 Acta Materialia, Inc.)

As a result, the electrochemically aligned collagen (ELAC) shows a similar degree of orientation to native tendons, although the characteristic D-bands in native fibrillar tendon collagen are not observed in ELAC (Fig. 3.4b) (Cheng et al., 2008). However, cross-linking with 1-ethyl-3-(3-dimethylaminopropyl) carbodiimide hydrochloride (EDC) in N-hydroxysuccinimide (NHS) solvent did generate the characteristic interchain spacing among the collagen triple helix, which was confirmed by X-ray diffraction (XRD) (Xie et al., 2021). On the other hand, cross-linking with glutaraldehyde was not able to restore these characteristic D-bands (Kew et al., 2011).

Electrochemically aligned collagen (ELAC) has several advantages when used for tissue engineering applications. As discussed previously, collagen presents natural motifs that facilitate cell adhesion, migration, and proliferation, resulting in positive biocompatibility ratings. Compared to traditional biodegradable polymeric materials, such as polylactic acid (PLA), ELAC scaffolds generated a tenfold increase in human umbilical vein endothelial cells (HUVECs) when cultured under the same conditions, supporting the claim that ELAC encourages cell adhesion, proliferation, and migration (Zhang et al., 2019). It was also observed that HUVECs seeded on ELAC scaffolds showed elongated and polarized morphology along the axial direction of the ELAC filaments (Zhang et al., 2019). Mesenchymal stem cells (MSCs) also presented a flattened and stretched morphology on ELAC scaffolds, suggesting good attachment to the surface (Kishore et al., 2012a). Tendon-derived fibroblasts also proliferated rapidly at a rate of 0.5 mm/day on ELAC scaffolds as shown in Fig. 3.4c (Cheng et al., 2008).

Not only does the natural motif facilitate cell adhesion, but the aligned topography of ELAC also mediates cell adhesion. It has been reported that the aligned morphology of ELAC improves the attachment of MSCs by a factor of 2 compared to a randomly oriented collagen gel (Kishore et al., 2012a, b). Even though the rate of proliferation of human MSCs was found to be slower on ELAC threads, this was not necessarily induced by the aligned feature of ELAC yarns. Given the initial high number of cells attached to ELAC, a proliferation plateau where the MSCs reached confluence is likely to have limited the long-term cell proliferation rate (Kishore et al., 2012a, b). Another potential reason is that MSCs prefer differentiation to proliferation on the aligned ELAC yarns (Kishore et al., 2012a, b). However, additional evidence is needed to confirm the tonogenic effect of ELAC scaffolds on MSC proliferation. The aligned morphology of the ELAC scaffold was also expected to polarize the morphology of HUVECs, whereas the randomly oriented electrospun collagen fibers induced a flat and less elongated morphology as observed by SEM (Zhang et al., 2019). However, in this study, the size of fibers on the ELAC scaffolds were on a different scale to the randomly electrospun fibers. The fibers on the ELAC scaffold were around 100 μm in diameter compared to about 1 μm for the electrospun fibers (Zhang et al., 2019). Therefore, further investigation is needed to support this result.

The highly aligned topography of the ELAC fibers was also found to signal cell differentiation. Kishore et al. showed that the aligned topography of the ELAC yarns induced MSC tenogenic differentiation compared to the random collagen gel, as supported by an increased expression of tendon specific markers (scleraxis and tenomodulin) and tendon ECM genes (tenascin-c and collagen-III), on the ELAC yarns (Kishore et al., 2012a, b). BMP-12 is a key growth factor that induces human MSC tenogenic differentiation. However, Kishore et al. found that the addition of BMP-12 did not further induce tenogenic differentiation of MSCs when they are cultured on ELAC scaffolds (Kishore et al., 2012a, b). This indicates that the ELAC alignment alone has a similar function to BMP-12 in inducing MSC tenogenic differentiation, which provides the possibility of using ELAC scaffolds without growth factors (Kishore et al., 2012a, b). Additionally, osteogenic differentiation of MSCs was suppressed on the ELAC scaffolds compared to the randomly aligned collagen gel, as observed by a reduced expression of a specific osteogenic marker, osteocalcin, on the ELAC yarns, which further supports a tenogenic effect of the ELAC scaffolds on MSCs (Kishore et al., 2012a, b). However, the specific mechanism of how the aligned ELAC morphology causes signal cell differentiation has yet to be explained.

ELAC monofilament yarns experience a faster rate of degradation compared to the synthetic polymeric materials such as poly(lactic acid) (PLA). ELAC generated from rat tail tendon collagen source was found to have a high tensile strength greater than 100 MPa and a tensile modulus over 200 MPa after 6 weeks of degradation in simulated physiological conditions (Kishore et al., 2012a, b). A braided ELAC scaffold applied to a rabbit's patellar tendon has been reported to degrade almost entirely within 8 months after implantation (Kishore et al., 2012a, b). A faster rate of degradation is associated with a faster resolution of the inflammatory response and therefore more rapid tissue regeneration and repair.

Due to the alignment and orientation of the collagen molecules, the ELAC monofilament yarn shows superior mechanical properties compared to the traditionally manufactured material such as a collagen gel or sponge, but has not yet achieved the mechanical performance, including the initial modulus and breaking stress, of the native tendon. However, the mechanical properties of the aligned cross-linked collagen filaments were observed to be superior to those of bleached tendons (Cheng et al., 2008). The ELAC monofilament yarns produced from rat tail tendon collagen had an ultimate tensile strength of 1.68 ± 0.32 N in the dry state (Zhang et al., 2019), which was sufficiently strong to allow them to be used for low-speed or manual weaving (Younesi et al., 2014), knitting (Zhang et al., 2019), and braiding operations (Kishore et al., 2012a, b). As a result, this bioactive ELAC scaffold can leverage its textile structure and be ready for commercial scale-up, since textile structures have the advantage of high strength, high flexibility, high porosity, a large surface area, and an anisotropic morphology.

A woven ELAC tendon scaffold has been reported by Younesi et al. and shows similar physical properties to native tendon tissue (Younesi et al., 2014). ELAC scaffolds upregulate tendon specific markers in mesenchymal stem cells (MSCs) in the absence of growth factors and secrete extracellular matrix (ECM) that encourages tendon repair in vitro (Younesi et al., 2014). Zhang et al. have reported on the design of a knitted ELAC scaffold as a tissue engineering vascular graft and have compared the mechanical properties to those of the saphenous vein (Zhang et al., 2019). The selection of an ELAC tissue scaffold compared to a pure polylactic acid scaffold also enhances the in vitro adhesion and proliferation of cardiosphere-derived cells (CDCs) (Xie et al., 2021) and human umbilical vein endothelial cells (HUVECs) (Zhang et al., 2019). In addition, Kishore et al. have evaluated braided ELAC scaffolds for tendon tissue engineering and have demonstrated the biocompatibility and biodegradability of the same ELAC yarns in vivo (Kishore et al., 2012a, b).

The strength and mechanical performance of collagen-based fibers still need further improvement so they can meet the various mechanical requirements for different tissue engineering applications. This requires optimization of the processing conditions, the level of cross-linking conditions, as well as the rate of degradation under dry and moist conditions.

During the processing of ELAC yarns, there are several variables that affect the collagen alignment, density, and tensile strength. First of all, a higher electrostatic force promotes collagen alignment (Dewle et al., 2020). As previously mentioned, the aggregation of collagen molecules occurs along a defined band where the isoelectric potential is focused and where a pH gradient has been established between electrodes (Akkus et al., 2018). Since type I collagen carries the same amount but the opposite charge near the two electrodes, it creates a drag or electrostatic force that increases the viscosity of the molecule (Cheng et al., 2008). It was observed that the viscosity near the anode was lower than near the cathode because there was a steeper pH change from the anode region to the isoelectric point than from the other side. This led to more collagen molecules being charged near the anode (Cheng et al., 2008). Consequently, greater repelling forces as well as faster migration of collagen molecules between the anode side and the isoelectric point result in an aligned collagen bundle near the anode (Cheng et al., 2008). Other factors that affect the bioelectrical conditions are having a suitable current density and electric field strength (Dewle et al., 2020). During the electrochemical induction, the current density impacts the amount of electric current flowing across the collagen solution, and the electric field strength accounts for the separation between the two electrodes. It has been observed that a suitable density and electric field strength for well alignment in the collagen range from 0.45 to 9.5 A/m^2 and 0.17 to 25 kV/m, respectively (Dewle et al., 2020). It has been found that changing the distance between the two electrodes modifies the charge distribution and the electric field strength across the entire range of collagen molecules, resulting in disruption of the fibril alignment (Dewle et al., 2020).

Cross-linking is another way to improve the mechanical properties of the collagen fibers. However, extensive cross-linking increases the stiffness of the protein and decreases elasticity. Cross-linking agents such as genipin and 1-ethyl-3-(-3-dimethylaminopropyl) carbodiimide hydrochloride (EDC) with N-hydroxysuccinimide (NHS) are the most used. Cross-linking ELAC using a 2% concentration of genipin and a 90% ethanol solvent to obtain a cross-linking degree of 65% significantly increased the wet ultimate tensile strength of the ELAC scaffold compared to the non-cross-linked sample (Uquillas et al., 2012). However, Xie et al. found that cross-linking with EDC/NHS did not directly improve the strength of the ELAC monofilaments but changed its morphology as well as slowed down its rate of degradation (Xie et al., 2021). It was shown that cross-linking with a ratio of collagen/EDC/NHS of 1:25:50 retarded the loss of mass and strength of the ELAC scaffold during an 8-week in vitro degradation assay in 0.01 M phosphate buffered saline (PBS) at 37 °C. A lower degree of cross-linking was achieved by a 2-h EDC/NHS treatment, which was better able to retard the strength loss compared to the higher level of cross-linking and the non-cross-linked samples. This prolonged the load-bearing period of the ELAC scaffold (Xie et al., 2021).

High-speed textile machines require robust, continuous, lubricated yarns for processing. To further adapt ELAC yarns for high-speed fabrication technologies, the ELAC yarns were blended with mechanically robust synthetic fibers. This allowed the manufacture of hybrid scaffolds using high-speed textile fabrication machinery, which will facilitate scale-up and the commercial production of ELAC-based biotextile scaffolds. Studies undertaken at North Carolina State University have demonstrated that plying ELAC monofilaments with PLA multifilament yarns can significantly increase the tensile strength of the hybrid yarn and facilitate the scaffold fabrication using high-speed textile machinery without compromising the excellent biological performance of the ELAC scaffold (Zhang et al., 2019). We have reported fabricating a woven PLA/ELAC scaffold for rotator cuff repair and a knitted PLA/ELAC scaffold for vascular tissue engineering using automated weaving and knitting machines (Zhang et al., 2019; Xie et al., 2022). Compared to the 100% pure PLA scaffold, the hybrid scaffold provided greater success in promoting adhesion and proliferation of tenogenic cells and endothelial cells (Xie, 2018; Zhang et al., 2019)

3.3 Collagen Yarn Extrusion

In order to extrude a continuous filament or yarn, wet spinning is widely used for processing collagen. Collagen solutions are prepared and extruded into a wet coagulation bath so as to produce a fibrillar structure during polymer coagulation and the formation of a continuous yarn or filament. Depending on the number of the filaments per yarn, there are two types of spun yarn: a monofilament, which only includes one filament per yarn, and a multifilament which includes more than one filament per yarn. The multifilament yarn is usually softer, more flexible, and less dense compared to a monofilament of the same size. However, a multifilament yarn is more difficult to spin because the finer filaments can be easily damaged or broken passing over a yarn guide's rough surface.

Extrusion is a process in which a molten material or a solution is forced through a die at a high production rate (several cm/s) to form a particular shape such as a bar, tube, or filament yarn. Collagen solution extrusion was first developed in the 1960s to create a new bio-suture for ophthalmic surgery. First a syringe is filled with the collagen solution, and then the syringe forces the solution through tubing into the collection or coagulation bath containing a fibrillogenic buffer. The process of self-assembly of the collagen fibers is either due to a change in pH (the acidic collagen solution is neutralized by the buffer) or due to a change in temperature. The diameter of collagen thread usually ranges from 100 μm to 1 mm in the wet state and around 20–100 μm after drying. Please note that these fiber dimensions are dependent on the syringe volume capacity as well as the inner diameter of the tubing (Rieu et al., 2017). However, the mechanical properties of the collagen yarn manufactured by this method are not as high as the native tendon. A study found that the Young's modulus of wet collagen yarn was about 4 MPa lower than that of native tendons, which range from 300 to 600 MPa depending on the tendon type (Johnson et al., 1994; Sano et al., 2013). Similar results were obtained from other research publications. Thus, a variety of cross-linking agents are widely used to improve the mechanical performance of extruded collagen yarns (Rieu et al., 2017) (Fig. 3.5).

The dry tensile strength of spun collagen has been improved by increasing the fibrillar alignment during the coagulation or polymer cross-linking steps. Koeck improved the wet-spinning process by extruding the collagen I/collagen III solution into a coagulation bath resulting in a self-assembled collagen monofilament without the need of additional cross-linkers. The monofilament size was up to 22 μm in diameter with a failure stress of 241 MPa and a Young's modulus of 5106 MPa. The monofilaments were then woven into flat and tubular shaped tapes on a narrow-width tape loom. This monofilament could be spun continuously at a max production rate of 34 m/h. Then the woven collagen scaffold was cocultured with myoblasts and fibroblasts so as to regenerate muscle-tendon junction tissue (Koeck et al., 2022). Similarly, the dry collagen filament strength was reported to lie in the range from 240 MPa (Siriwardane et al., 2014) to 383 MPa (Haynl et al., 2016). Tonndorf's wet-spinning experiment spun multifilament yarns (six filaments per yarn) with each filament having a diameter of 80 μm. The tensile strength was 169 MPa and the Young's modulus was 3534 MPa in the dry state. In comparison, the tensile strength was 40 MPa, and the Young's modulus was 281 MPa under wet conditions (Tonndorf et al., 2020).

Hoogenkamp et al. created a new technique to collect the collagen yarns from an extrusion system equipped with counterrotating cones. Basically, a mixed collagen gel was loaded into the extrusion system, and a metering pump forced the gel through the die at a constant pressure. After passing through the metering pump, the collagen suspension traveled to the extrusion head in between the counterrotating inner and outer cones and was finally extruded through a circular die and deposited on a mandrel (Fig. 3.6). By varying the inner and outer cone speeds, the collagen

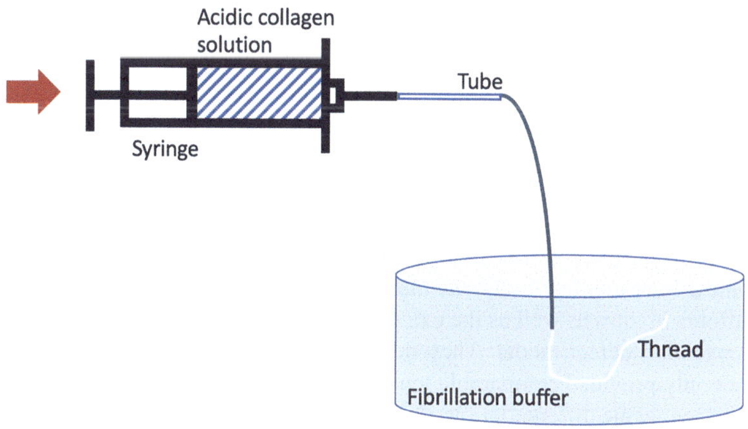

Fig. 3.5 Schematic diagram showing the results of extruding collagen solution

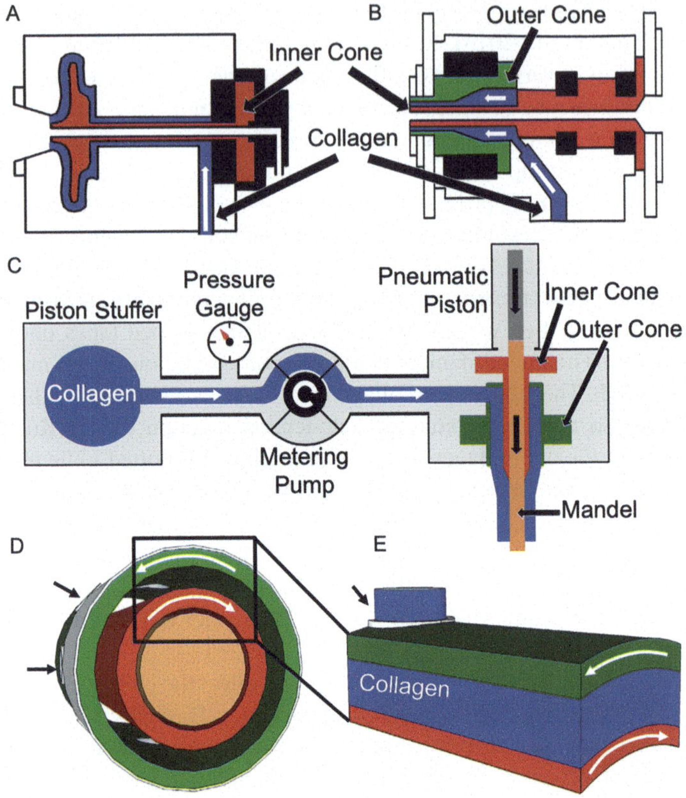

Fig. 3.6 Overview of the system for extrusion technology. (**a**) Single cone extruder design. (**b**) Counterrotating cone extruder design. The collagen gel (blue) is extruded between the inner cone (red) and the outer cone (green) which are the rotating parts. White arrows indicate the direction of collagen flow. (**c**) Diagram of the collagen delivery system. (**d**) and (**e**) Cross-sectional views of the extrusion head. (Reprinted with permission from Hoogenkamp et al., 2015. Copyright © 2014 Acta Materialia Inc.)

showed different orientations with varying degrees of elasticity and toughness (Hoogenkamp et al., 2015).

Yang et al. further modified the counterrotating cone extrusion approach. They produced well-aligned collagen fiber scaffolds by increasing the cone speed. A higher cone speed generated stronger shear forces across the wall of the tube and resulted in a narrower distribution of collagen fibers. The authors evaluated the tenogenic differentiation of rat bone marrow mesenchymal stem cells (rBMSCs) on the scaffolds in vitro as well as the extent of tendon regeneration in vivo using a rat Achilles tendon defect model. They demonstrated that the aligned collagen scaffolds not only provided comparable tensile strength to native tendon tissue but also promoted tenogenic differentiation of rBMSCs, making them a suitable scaffold for tendon tissue regeneration (Yang et al., 2019).

Nakayama et al. developed a shear-based fibrillogenesis technique as shown in Fig. 3.7. A collagen solution was concentrated to a liquid crystal state and extruded onto a plastic surface. After the solvent had evaporated, a film of aligned collagen nanofibers with a diameter of 30 nm was formed as shown in Fig. 3.7a. The scaffold was further cross-linked using a dehydrothermal treatment (DHT) at 110 °C under 50 mTorr vacuum for 48 h (Nakayama et al., 2015). The collagen yarns produced by this method contained collagen fibrils with a diameter of about 30 nm, and the crimp lays perpendicular to the axial direction as typically seen in natural collagen (Fig. 3.7b, c) (Nakayama et al., 2015). This extruded collagen scaffold was found to support a monolayer of endothelial cell migration that extended from the scaffold to neighboring tissues (Nakayama et al., 2015). The parallel alignment of the scaffold, as observed by Nakayama et al., also supported myotubular development by increasing the alignment of the myotubes and developing an aligned and stratified morphology. The aligned graft was also found to improve the contractile function of the film (Nakayama et al., 2019).

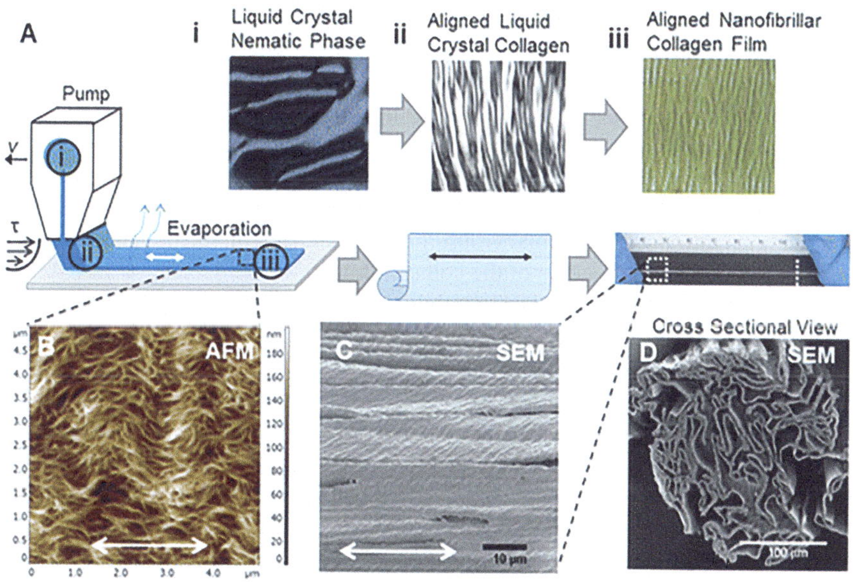

Fig. 3.7 (**a**) Liquid crystal collagen was pumped onto a plastic surface and formed an aligned nanofibrillar collagen film after solvent evaporation. The film self-assembled into a thread, which contains (**b**) a crimped structure and (**c**) an aligned morphology like native collagen. (**d**) A cross-sectional view shows the porous structure. (Reprinted with permission from Nakayama et al., 2015. Copyright © 2015 American Chemical Society)

3.4 Collagen Modification

Collagen fibers are known to experience erosion and degradation when exposed in vivo to body fluids and enzymes such as collagenase. This results in inferior mechanical properties compared to collagen in the natural state whose structure and level of cross-linking has been destroyed during extraction (Charulatha & Rajaram, 2003; Delgado et al., 2014). Considering the importance of collagen in tissue engineering and regenerative medicine, it is essential to not only ensure its biocompatibility but also to maintain its structural integrity and mechanical performance so as to meet the requirements for further fabrication and clinical applications.

As mentioned above, type I collagen is composed of three alpha chains: two alpha1 chains and one alpha2 chain. Within each chain, the amino acid residues are connected by covalent bonds as well as non-covalent bonds, such as hydrogen bonds, so as to maintain the integrity of the protein structure. There are many reactive groups within the collagen molecule that provide the possibility for surface modification. First, the many amino acid residues, such as glycine and hydroxyproline, provide hydrogen bonding donor sites, while the carbonyl groups (C=O) in the peptide chains provide hydrogen bonding acceptance sites. This results in an abundance of hydrogen bonds. Free carboxyl groups in the collagen molecule can react with active amino groups, and free ε-NH_2 groups in the lysine residues can react with aldehydes and compounds containing the aldehyde group (Yu et al., 2016).

During the last decade, many modifications to collagen chemistry have been proposed and evaluated with the objective of making collagen directly available for various clinical applications. Cross-linking is one of the most widely studied modifications, since it introduces physical or chemical links between collagen molecules or between exogenous molecules and collagen (Gu et al., 2019). Physical methods improve the stability of the collagen structure by applying energy that activates more radicals which then form new intermolecular bonds, while chemical modification methods introduce functional groups into the collagen molecules and thus alter the physicochemical properties of the collagen fibers.

Dehydrothermal treatment (DHT) stabilizes the collagen structure by promoting the formation of cross-linking amide bonds between the free carboxyl and amine groups within the triple helices. DHT treatment first exposes the samples to different temperatures (110 °C, 150 °C, or 200 °C) for 24–48 h, before cooling them down to room temperature in a vacuum furnace. The cross-linking and denaturation of the collagen occur simultaneously during the process. The effect of both modifications increases with increasing temperature, especially the degree of denaturation, which increases sharply when the temperature exceeds 150 °C (Haugh et al., 2009; Toba et al., 2002; Wessels & Pretorius, 2008). Because it has been observed that exposure time does not affect the cross-linking density, it was recommended that all collagen samples be treated at 110 °C for 24 h (Haugh et al., 2006). After the DHT treatment, the tensile properties of the collagen scaffolds increased, with the tensile strength reaching 3.2 MPa, an 1.8-fold increase for the scaffolds cross-linked at 200 °C compared to the non-cross-linked collagen scaffolds. And when exposed to bacterial collagenase, the cross-linked collagen scaffolds showed better resistance to degradation (68% degraded) compared to the non-cross-linked ones (80% degraded) after 7 days (Hu et al., 2013).

The use of ultraviolet (UV) irradiation is another method for cross-linking collagen. The principle of this physical method is to induce the formation of free radicals in certain amino acids such as tyrosine and phenylalanine which contain double bonds and aromatic rings. Following UV irradiation, the free radicals produced will generate intermolecular covalent bonds (Wihodo & Moraru, 2013). The tensile properties of collagen fibers including the ultimate tensile strength, the Young's modulus, and ultimate strain at break have been observed to increase significantly after 15 min of UV irradiation (Weadock et al., 1995). However, the mechanical properties and enzymatic stability of the collagen structure cross-linked with UV light alone are not sufficiently stable, and as a result, it has been suggested that the UV irradiation treatment be combined with other cross-linking methods (Yu et al., 2016).

In addition to DHT treatment and UV irradiation, plasma modification is another surface modification technique for cross-linking collagen. The purpose of plasma modification is to introduce amines and other functionalities, such as carboxyl and hydroxyl groups, to the collagen surface by adding air, O_2, and NH_3 to the inert gases (Ar, He) when the plasma is generated. This will modify the collagen surface with the introduction of covalent bonds (Taraballi et al., 2013; Yu et al., 2016).

The physical modification technologies described above are relatively easy to operate, safe, and low cost, but the cross-linking efficiency is not comparable to chemical cross-linking. Common chemicals used to cross-link collagen are glutaraldehyde, 1-ethyl-3-(3-dimethyl aminopropyl) carbodiimide (EDC), and genipin. Glutaraldehyde is the oldest chemical used for chemically cross-linking collagen. It introduces exogenous cross-links between itself and the collagen molecules by reacting with certain amino groups such as the lysine and hydroxylysine residues within the collagen fibrils (Casali et al., 2018; Duan et al., 2018). One advantage of using glutaraldehyde is that it will protect collagen from being degraded by collagenase. However, there are still concerns about using glutaraldehyde to cross-link collagen, mainly because of observed adverse cellular effects associated with the inhibition of fibroblast growth, even after removing any unreacted glutaraldehyde (Haugh et al., 2011; Xu et al., 2013).

The use of 1-ethyl-3-(3-dimethylaminopropyl) carbodiimide/N--hydroxysuccinimide (EDC/NHS) dendrimers has been investigated as a potential cross-linking agent to replacement glutaraldehyde. With the presence of EDC/NHS, the carboxylic acid functionality of the glutamic acid and aspartic acid residues on one amino acid sidechain are able to react with the primary amine on an adjacent amino acid through a condensation reaction (Gu et al., 2019). Contrary to cross-linking with glutaraldehyde, this method does not introduce any cytotoxic activity into the collagen molecule. NHS is used as the catalyst, and excess EDC can be removed from the collagen by a sufficient number of wash cycles (Ahn et al., 2013). Yu cross-linked collagen yarns with EDC/NHS for different periods of time (2 and 17 h) and compared the degradation performance for these low and high cross-linked samples as well as for a non-cross-linked collagen control over a period of 8 weeks. The results showed that the initial maximum load for the non-cross-linked collagen yarns was significantly lower than for the cross-linked samples. There was no significant difference in mass loss between the low and high cross-linked yarns which experienced an approximately 68% and 84% decrease in their maximum load between week 6 and week 8, respectively. The non-cross-linked collagen yarns were almost completely degraded at week 8 (Xie et al., 2021).

Genipin is a natural cross-linking agent for collagen. It is derived from geniposide, found in the fruit of the *Genipa americana* plant. It has been found that genipin is approximately 10,000 times less cytotoxic than glutaraldehyde and about 5000 times better than glutaraldehyde in promoting cell attachment and proliferation (Cauich-Rodriguez et al., 1996).

Mekhail et al. cross-linked electrospun type I collagen fibers with genipin and evaluated their biocompatibility. They found that electrospun collagen fibers were able to maintain their morphology after being cross-linked with genipin in an aqueous environment (Mekhail et al., 2011).

The resistance to enzymatic degradation of collagen after cross-linking with genipin has also been demonstrated. Zhang et al. reported that the degradation resistance of collagen scaffolds increased significantly when cross-linked with 0.3% genipin at 37 °C (Zhang et al., 2014). Bi et al. studied the effect of different cross-linking conditions on the resulting properties of collagen scaffolds and found that as the genipin concentration, the cross-linking time, and cross-linking temperature increased, so the degree of cross-linking also increased. However, at higher genipin concentrations, the observed rate of degradation was slower (Bi et al., 2011).

3.5 Surface Modification

Collagen is the most common protein found in the extracellular matrix (ECM) of most, if not all, living tissues. Thus, when coated onto a hydrophobic biomaterial, it not only increases the hydrophilicity but significantly improves the biocompatibility of the synthetic biomaterial surface, along with retaining its bulk physical and mechanical properties (von der Mark et al., 2010). So applying a collagen coating to a polymeric scaffold has become a well-established procedure for improving the functionality of a hydrophobic surface and increasing cell attachment. Heinemann et al. observed a significant improvement ($p < 0.001$) in proliferation and differentiation of osteoblasts when cultured on a chitosan fiber-based nonwoven scaffold which was immobilized with a bovine collagen type I coating when compared to the uncoated fibrous scaffold and an organically modified hydroxyapatite scaffold control (Heinemann et al., 2019). The process of coating a fibrous scaffold with a collagen solution not only increases its biocompatibility in terms of cell attachment but also maintains its porous structure due to the even distribution of the coating along the surface of individual fibers which does not block or seal the pores that support cell penetration and growth. However, this may not apply to nanofiber electrospun scaffolds, as the web of 300-nm-long collagen fibrils will contain nanosized pores which are too small to allow the human bone marrow stromal cells to penetrate into the scaffold (Lodish et al., 2000) (Fig. 3.8).

However, the effectiveness of the surface modification may differ depending on the technique used to coat or immobilize the collagen solution on the scaffold. In a study by Sogorkova et al., narrow-width knitted scaffolds were prepared from wet-spun palmitoyl-hyaluronan derived (palmHA) multifilament yarns. These scaffolds were coated with five different proteins including fibronectin from human plasma, fibrinogen from bovine plasma, laminin from a murine sarcoma basement membrane, methacrylated gelatin from porcine skin, and collagen IV obtained from human cell culture. To coat the scaffolds, they were first impregnated with the saline solution of the protein, followed by UV cross-linking and stabilization with

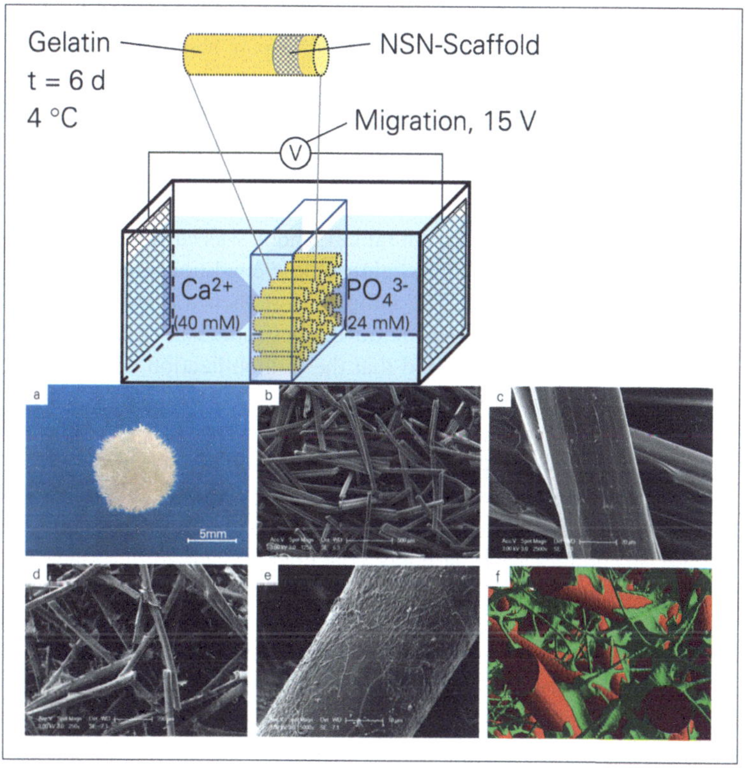

Fig. 3.8 Schematic diagram of a calcium phosphate modification and mineralization technique involving a double migration method with a polyacrylic barrier on a net-shape-nonwoven (NSN) chitosan scaffold surface (top). Image (**a**) is a photograph of the nonwoven scaffold; (**b**) and (**c**) are SEM images of the scaffold before collagen coating; (**d**) SEM image of the nonwoven scaffold treated with collagen; (**e**) SEM image of collagen fibrils attached to the surface of an individual filament; (**f**) visualization of stained scaffold showing red chitosan fiber component and green collagen component. (Reprinted under Creative Commons Attribution 3.0 license from Heinemann et al., 2019)

2-hydroxy-4′-(2-hydroxyethoxy)-2-methylpropiophenone, a saline solution wash and air drying. However, the authors found that the use of a dip coating method to apply collagen IV generated the least uniform surface coating (Sogorkova et al., 2018).

Immobilization of collagen by grafting on the surface of a textile biomaterial is advantageous over simple dip coating or impregnation, because grafting is stable and durable, and has been observed to retain its surface properties even after harsh changes in temperature or pH (Cheng & Teoh, 2004).

Grafting of collagen on a biomaterial surface can be carried out in two steps: first surface activation and then graft polymerization. Surface activation is required to create functional groups on the backbone chain of the polymeric material after which the desired protein molecule is bonded to the functional group on the surface (Cheng & Teoh, 2004). For example, Park et al. used collagen immobilization to improve the biocompatibility of a poly(ε-caprolactone) (PCL) ultrathin nanofiber membrane toward endothelial cells. The electrospun nanofiber membrane was treated with 0.1 M sodium hydroxide at room temperature, so as to make the originally hydrophobic material hydrophilic. The hydrolysis reaction altered the ester chains on the PCL backbone into carboxylic (R-COOH) groups. This was followed by treatment with a 1:1 solution of 100 mM EDC and 100 mM sulfo-NHS in a 0.625 M 2-(N-morpholino) ethanesulfonic acid (MES) buffer. The subsequent immobilization of collagen type I from rat tail tendon was carried out at room temperature, in a humid chamber, under slightly acidic conditions made by dissolving collagen in 0.625 M MES buffer. The samples were then washed three times with PBS containing sodium phosphate to remove any unattached collagen and to neutralize any residual acid. It was found that the immobilization of collagen on the surface through grafting generated a nanofiber membrane with a more uniform surface which resulted in a longer period of cell viability compared to both the uncoated membrane and the membrane physically coated with collagen (Park et al., 2019).

Plasma surface modification (PSM) is one of the more frequently studied and attractive treatments for surface activation of hydrophobic polymeric biomaterials. It has several advantages such as being cost-effective, environmentally friendly, reproducible, and easy to implement. Plasma is an ionized gas, the fourth state of matter, a mixture of positively and negatively charged ions and molecules. Plasma gas is developed by applying an electric current between two electrodes inside a chamber filled with a vacuum or an inert gas. At low vacuum pressure, all atoms and molecules inside the chamber are ionized, and a glow discharge plasma is created at room temperature. Radio frequency glow discharge (RFGD), corona discharge, and atmospheric arc plasma are three of the common subtypes of gas plasma. The charged ions are bombarded onto a biomaterial surface, which transfers energy from the chamber to the surface of the biomaterial. In order to functionalize the surface, inert gases such as He, Ne, and Ar are used to introduce free radicals and create active sites on the biomaterial's surface. Reactive gases, such as oxygen, ammonia, and fluorinated compounds, are introduced to create functional groups, such as peroxides, hydroperoxides, and amino groups on the surface. Subsequently at a later stage, graft polymerization is performed to graft the desired protein molecules onto the scaffold's surface (Solouk et al., 2011). In recent years, plasma surface modification has been used in the fields of tissue engineering, wound healing, and sterilization of medical devices (Solouk et al., 2011) (Fig. 3.9).

Cheng et al. reported a process for modifying ultrathin polycaprolactone (PCL) films fabricated by solvent casting and biaxial stretching. They used this strategy to improve the biocompatibility and hydrophilicity of PCL films for tissue engineering applications. For collagen immobilization, they first activated the surface of the films by Ar plasma pretreatment in a quartz cylindrical-type glow discharge cell. The pretreated PCL film was then exposed to UV light which initiated an acrylic acid (AAc) graft polymerization reaction which took 30 min at room temperature, followed by collagen immobilization. Before treating the samples with calf skin type I collagen solution in PBS, the authors used 1-ethyl-3-(3-dimethylaminopropyl) carbodiimide (EDC) to improve the hydrophilicity of the PCL film, as reflected by the reduction in the water contact angle, as well as improve the biocompatibility in terms of the attachment and proliferation of human dermal fibroblasts and myoblasts (Cheng & Teoh, 2004).

Suntornnond et al. analyzed and optimized the parameters of gas plasma treatment, used to activate the surface of PCL membranes prior to graft collagen immobilization. The PCL membranes were fabricated from PCL powder using an organic, solvent-free suspension heating system. Various parameters, such as the type of gas (argon and oxygen), gas flow rate (200 and 100 cm^3/min), and treatment time (180s and 120s), were included in a 2k factorial design of experiment. The hydrophobicity

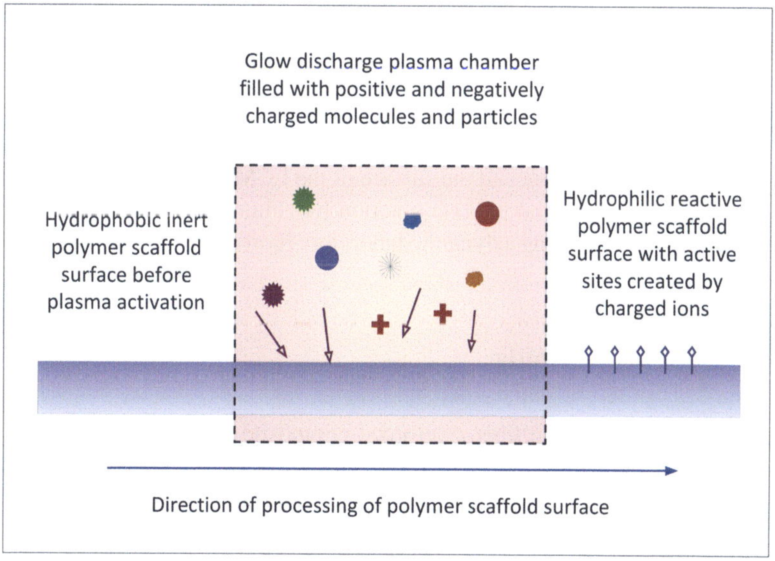

Fig. 3.9 Schematic diagram of hydrophobic polymer scaffold surface being modified by exposure to glow discharge plasma treatment

of the PCL membranes' surface was reduced from 70 to around 40 degrees as measured by a contact angle goniometer. The experimental results confirmed that the reduction in hydrophobicity could be achieved on the PCL membranes by plasma treatment using argon gas, a gas flow rate of 100 cm^3/min for a treatment time of 120 s. The plasma pretreatment with argon introduced functional groups onto the membranes' surface, which were subsequently immersed in a solution containing 0.11 g/ml acrylic acid and vitamin B. After exposure to EDC in an aqueous solution for 1 h, the samples were treated with 0.5 g/ml type I collagen solution for 5 h at 4 °C. There was significant improvement in the biocompatibility of the membranes toward L929 fibroblasts in terms of cell proliferation, due to plasma pretreatment and collagen immobilization (Suntornnond et al., 2016).

Liu utilized plasma treatment for surface modification of a polypropylene (PP) hernia repair mesh and functionalized it with maleic acid prior to grafting it first with genipin and then with collagen. The plasma treatment and collagen coating were performed to improve the biocompatibility of the surface and reduce the risk of infection and rejection after implantation. The hernia repair mesh had improved biocompatibility in terms of improved cell attachment and proliferation using neonatal human dermal fibroblasts (Liu, 2019). The genipin molecule served as a spacer between the polymer backbone chain and the collagen coating material. Using a spacer molecule like genipin is advantageous because it allows the collagen chains to acquire their natural conformation while being attached to the scaffold. This not only improves the stability of the coating but also provides a surface that closely mimics the natural ECM.

Ramakrishna also incorporated collagen immobilization using genipin as a spacer molecule to bioactivate scaffolds and provide a biomimicking surface for tendon-bone regeneration. The fabricated device was composed of a biphasic double layered tubular scaffold braided and knitted from PLA yarns of different cross-sectional shapes. The scaffolds were then surface activated in an oxygen-rich atmospheric radio frequency plasma and functionalized with maleic acid. After coating with collagen, the scaffolds were then coated with hydroxyapatite to form a structural gradient that mimicked the surface of the ECM at a tendon-bone junction. The study regenerated a tendon-bone junction by culturing tenocytes in vitro with mechanical stimulation using a dynamic laboratory bioreactor (Ramakrishna, 2019).

3.6 Biotextile Scaffolds

Biotextile scaffolds include fiber-reinforced composite structures as well as scaffolds fabricated from fibers and yarns using textile processing equipment that are expected to function in a biological environment (King et al., 2019). The textile fabrication processes include weaving, knitting, braiding, flocking, embroidery, nonwoven production, and electrospinning in both 2D and 3D.

Native ECM is a composite material where soft proteins are reinforced by fibrous proteins to form a structure with various functions (Fernández-Colino &

Jockenhoevel, 2020). For example, bone tissue offers high stiffness and hardness, while tendon tissues present excellent extensibility and elasticity. In cartilage, the collagen fibrillar structure limits the swelling of the proteoglycan matrix. The tension between the matrix and the collagen fibrils counteracts the swelling force, which therefore prevents the proteoglycan matrix from swelling and stiffens the cartilage's mechanical behavior.

Hydrogels made from collagen and its derivatives, such as thermo-cross-linkable collagen gel, gelatin, and gelatin methacryloyl, have been studied extensively and applied to tissue engineering end uses with excellent biological performance. They have high water content and present biological motifs that facilitate cell adhesion, migration, proliferation, and differentiation. Integrin is an important cell transmembrane receptor that recognizes and interacts with extracellular matrix proteins, such as collagen. Cha et al. found that integrin plays an important role in resolving macrophage inflammation, which is associated with reduced scar tissue formation and accelerated tissue regeneration (Cha et al., 2017). Collagen derived hydrogel is designed to provide cells with a microenvironment that resembles native ECM with appropriate biological motifs, a three-dimensional (3D) structure, physiologically relevant stiffness, and a hydrated environment (Visser et al., 2015). However, a pure gel made solely from collagen or its derivatives usually has inferior mechanical properties due to insufficient restoration of the molecular helix structure and poor alignment (Lin & Gu, 2015). The swelling behavior of the pure collagen hydrogel and the contraction by the cells in culture also introduce structural instability which limits its clinical use (McCullen et al., 2010).

A composite scaffold made from a gel matrix reinforced by fibers or fabrics has been widely studied and found to mimic the native ECM structure. The composite construction provides a balance between the need of bioactivity and mechanical durability. One of the earliest tissue-engineered vascular grafts was made from a collagen gel reinforced by a Dacron polyester sleeve (Niklason & Lawson, 2020). Fiber-reinforced composites typically show superior mechanical properties compared to the gel alone, which generally means that a composite structure is required for load-bearing applications (Visser et al., 2015). At the same time, the high water content and the cell-binding motifs of collagen gel restrict its biomedical application (Yoon et al., 2019). Yoon et al. reported that by embedding layers of electrospun nanofiber mats in between the printed gel layers, the dimensional stability and mechanical robustness of the construct were significantly improved (Yoon et al., 2019).

Commonly seen reinforcing components include short fibers, continuous filaments, monofilament and multifilament yarn, two-dimensional textile structures (whether woven, knitted, braided, or nonwoven), and additive manufactured constructs. Different reinforcements can be selected and designed based on specific mechanical requirements of the target tissue.

Short fibers can significantly improve the mechanical properties of a gel even with only a small number of fibers distributed in it. This is due to the fiber's high surface area-to-volume ratio. The interstitial adhesion between fiber and gel enhances the mechanical performance of the composite structure. The strength and strain of the composite depends on the adhesion between the reinforcing fibers and the gel, as well as on the strength and strain of the fiber or yarn (Agrawal et al., 2013). The fiber diameter, fiber volume, type of raw material, and the interaction between the fiber reinforcement and the matrix material are the key factors that determine the final mechanical properties of the composite structure.

Two-dimensional textile structures, including electrospinning, weaving, knitting, braiding, flocking, and embroidery, have certain key features such as being strong and tough yet flexible and extensible and easy to process and scale-up for industrial manufacturers. All of these features make textile technologies commercially relevant for medical device manufacturers.

A woven structure has long been selected for the fabrication of commercial synthetic vascular grafts from Dacron® polyester yarns. Weaving involves interlacing two sets of yarns perpendicular to each other on a weaving machine or loom (King et al., 2019). The fabric direction is the warp direction, and the perpendicular direction is called weft direction. Warp yarns must have sufficient strength to withstand the stresses during the weaving process, whereas the mechanical strength requirement for the weft yarns is much lower, which allows the utilization of a wide variety of potential materials with relatively poor mechanical properties but with attractive functionality such as good biocompatibility and biodegradability (Xie et al., 2022). Most woven structures are stable and durable, and for this reason, they are selected for applications that require a high level of structural integrity and stability.

Knitted structures are fabricated by interlooping yarns through one or more set of hooked needles (King et al., 2019). They provide good compliance, stretchability, and elasticity to the scaffold fabric, and thus they are suitable for applications that require extensibility and elasticity, that is to say, "stretch and recovery" (Zhang et al., 2023b). However, there are certain inherent disadvantages when using a knitted structure. For example, they have poor dimensional stability and tend to fray and unravel at the unsecured end. You may have experienced this personally when your knitted track suit bags at the knees and elbows! If a yarn breaks, this can lead to a "run" (in hosiery) or a hole in the knitted structure. Their porosity is extremely high, often in excess of 95%, which means they may not provide enough cover or opacity when worn as clothing. As a result, the yarns at the edge are often thermally sealed (Zhang et al., 2021) and the fabric is heatset (Xie et al., 2016) so as to prevent distortion, deformation, or unraveling of the knitted scaffold.

Braiding involves interlacing three or more threads or yarns with each other to form a fabric (King et al., 2019). If the braiding machine has an even number of bobbins or yarns, half of them will move clockwise and the other half will move counterclockwise, producing a tubular structure similar to a May pole. However, if the braiding machine has an odd number of bobbins or yarns, then it will produce a flat braid. The main challenge when designing any braided structure is how to seal the ends so the structure does not fray and unravel. Braided structures are often used in medical applications as sutures and stents because of their structural flexibility and mechanical robustness and durability in the longitudinal direction (Akbari et al., 2016). However, the braided structure typically has free ends which need to be secured by interlacing yarns which ensure its structural integrity and stability.

Fernández-Colino et al. designed a small caliber tubular scaffold for in situ vascular tissue engineering. A warp-knitted polyvinylidene fluoride (PVDF) fabric was embedded in an elastin-like recombinamer (ELR) to form a PVDF-reinforced ELR tubular scaffold. The incorporation of knitted PVDF reinforcement made it possible for the graft to have sufficient burst pressure. The composite scaffold was further covered by an electrospun PCL outer sheath to tune the scaffold's compliance so as to match different implantation targets. ELR matrix supported the formation of an intact endothelium and ECM protein deposition by SMCs which barely showed any platelet activation. (Fernández-Colino et al., 2019). Since this study was limited to an in vitro approach, this topic needs to be studied further in a clinically relevant animal model. Having said that, the current study was able to demonstrate the advantage of a textile-reinforced composite scaffold in balancing the mechanical and biological performance, which is hard to achieve with a single-component scaffold (Fig. 3.10).

We have designed a circular-knitted vascular graft composited with gelatin methacrylate (GelMA) hydrogel (Zhang et al., 2023a, b). The circular-knitted textile reinforcement provides primary mechanical robustness to the vascular graft (Fig. 3.11). The textile is sealed with GelMA hydrogel material, resulting in reduced pore size of the structure and increased endothelial cell proliferation in vitro (Zhang et al., 2023b).

Nonwoven fabrics are another important category of a two-dimensional textile structure. It is fabricated by forming a web by laying down short staple fibers or continuous filaments in either a random or an aligned configuration. The nonwoven fiber-forming techniques include carding, air lay, wet lay, spun-bond, melt-blown, and more recently electrospinning techniques. In a second step, the fibers in the web need to be bonded together, and there are a series of different bonding techniques depending on whether the fibers are thermoplastic or not. They include needle punching, hydro-entanglement, thermal bonding, stitch bonding, and chemical (adhesive) bonding methods.

Electrospinning is another versatile textile technology which has the potential to be involved in a broad range of applications, which were discussed in detail in Sect. 3.1. The main advantage of using an electrospun web or mat is the nanosized filaments that are attenuated during the electrospinning process. This provides high surface area which has a similar topography to natural collagen's extracellular matrix.

Yoon et al. designed a PCL fiber-reinforced alginate hydrogel construct for soft tissue engineering (Yoon et al., 2019). The PCL fibers were used as an interstitial layer in between the 3D-printed alginate gel. Embedding electrospun fibers enhanced and stabilized the shape of the scaffold as well as improving the mechanical robustness of the hydrogel construct. The reinforced composite recovered almost completely from a compression test applied in the thickness direction, whereas the unreinforced pure hydrogel was irreversibly compressed.

The electrospun mat also prevented the woodpile-patterned hydrogel from causing the scaffold to collapse by providing lateral permeability. On the other hand, the interconnecting porous structure of the electrospun mat ensured there was also

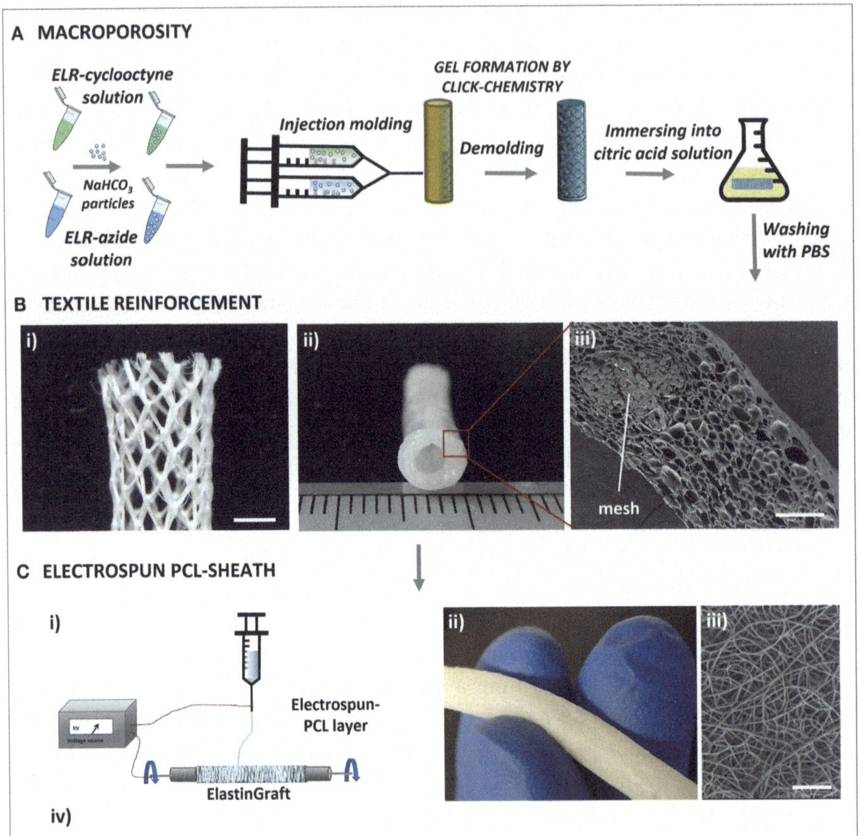

	PCL sheath thickness (μm)
ElastinGraft_0	0
ElastinGraft_eS	10 ± 4
ElastinGraft_eM	30 ± 6
ElastinGraft_eL	232 ± 26

Fig. 3.10 Textile-reinforced vascular scaffold. (Reprinted with permission from (Fernández-Colino et al., 2019) under Creative Commons Attribution License. Copyright © 2019 Fernández-Colino, Wolf, Rütten, Schmitz-Rode, Rodríguez-Cabello, Jockenhoevel and Mela)

vertical permeability through the thickness of the scaffold. Ultimately, the design of the composite enhanced structural permeability which facilitated mass transfer and improved cell viability (Yoon et al., 2019).

In addition, Yoon et al. suggested that the oxygen plasma treatment on the electrospun web significantly improved the hydrophilicity of the fibers and enhanced the intimate adhesion between the electrospun fibers and the gel. As a result of this improved adhesion, multilayered composite constructs were successfully fabricated from alginate hydrogel embedded with an interstitial electrospun PCL fiber mat.

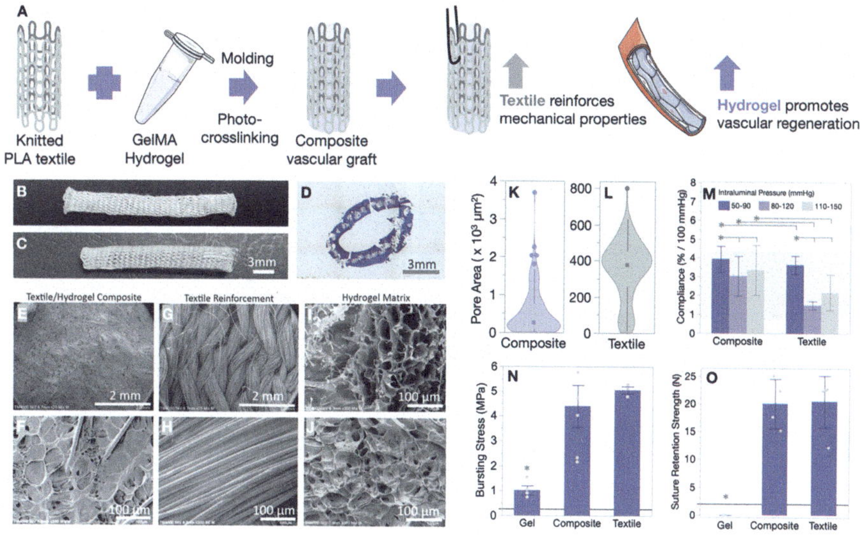

Fig. 3.11 Composite vascular scaffold reinforced with knitted textile. (**a**) Schematic of vascular graft design and fabrication. (**b**) Circular-knitted textile reinforcement. (**c**) Textile-GelMA composite vascular graft. (**d**) Hematoxylin and eosin staining of the cross section of the composite vascular graft. (**e–j**) Scanning electron microscopies of vascular graft surfaces of composite graft, textile, and hydrogel, respectively. (**k** and **l**) Pore area distribution of (**k**) composite graft and (**l**) knitted textile graft. (**m**) Compliance of composite and textile vascular grafts. (**n**) Bursting strength and (**o**) suture retention strength of vascular grafts made from composite, GelMA alone, and textile. (Reprinted with permission from Zhang et al., 2023b. Copyright © 2023 The Royal Society of Chemistry)

However, the link between the improvement in the hydrophilicity and the ultimate performance of the construct has not yet been studied (Yoon et al., 2019).

Solution blown spinning (SBS) also generates a controlled arrangement of fibers within a nonwoven fabric similar to melt blown spinning (Vasireddi et al., 2019). Akentjew et al. designed a vascular graft using a gelatin methacryloyl/alginate (GEAL) hydrogel reinforced with polycaprolactone (PCL) fibers to mimic the multilayer, fibrous arrangement of the native ECM in coronary arteries (Akentjew et al., 2019). The fibers were deposited at specific angles using SBS technology and collected by a dip-spinning apparatus to form a tubular structure. At the same time, the apparatus moved vertically to immerse the fibers in the hydrogel. After multiple cycles, a multilayer PCL fiber-reinforced GEAL composite vascular graft was built. And from a mechanical point-of-view, the graft successfully mimicked the stress-strain curve of the human coronary artery (Akentjew et al., 2019).

In addition to traditional textile technologies, textile inspired structures are also fabricated by additive manufacturing technologies such as 3D printing and direct jet deposition. Kong et al. developed a composite scaffold by first fabricating a grid-like matrix using direct jet deposition technology and then infusing the scaffold with a GelMA hydrogel matrix. The composite scaffold showed enhanced mechanical properties compared to pure hydrogel (Kong et al., 2020) (Fig. 3.12).

Kong et al. found that the fiber spacing is inversely proportional to the mechanical properties of the scaffold, indicating that one can design the spacing of the fibers

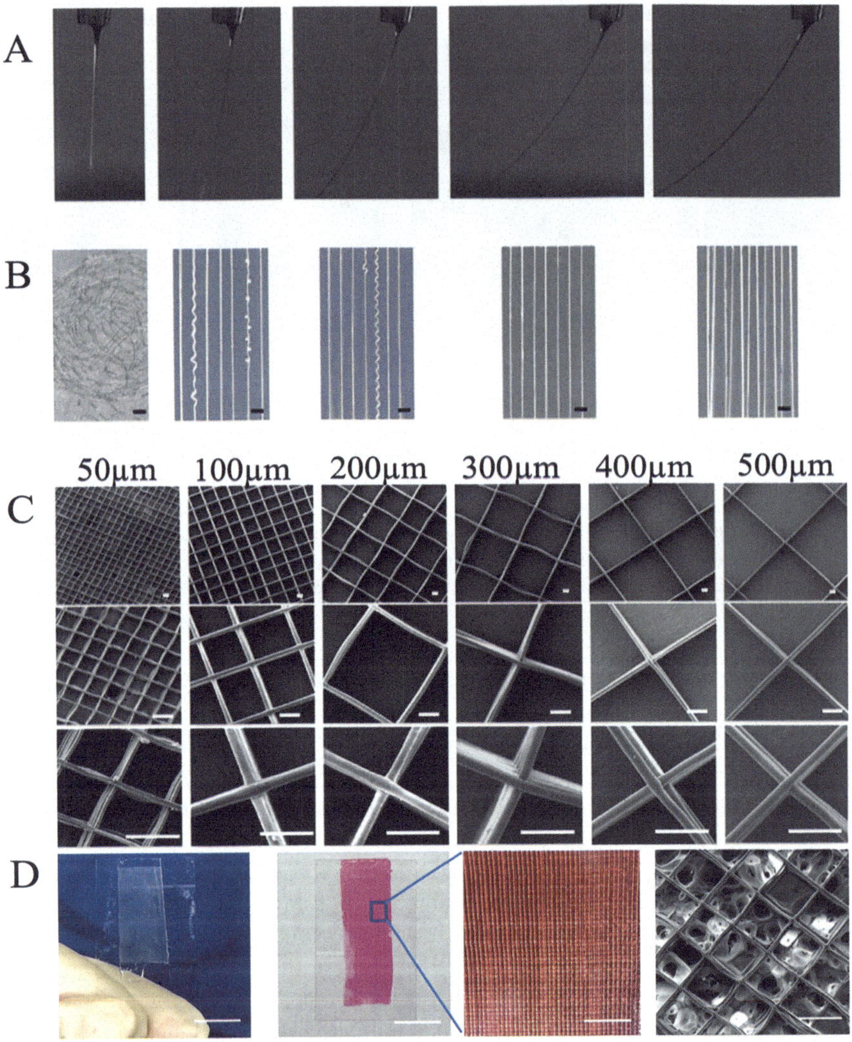

Fig. 3.12 Fiber-reinforced corneal equivalent. (**a**) Images of stable jet deposition at different moving speeds of the substrate. (**b**) The resulting fiber alignment (scale bar = 100 μm). (**c**) SEM images of the grid scaffolds with fiber spacing from 50 to 500 μm under different magnifications (scale bar = 50 μm). (**d**) Macroscopic and microscopic morphology of the 100 μm spaced fiber-reinforced hydrogel (scale bar in macroscopic images = 10 mm, in the enlarged image = 1 mm, and in the SEM image = 100 μm). (Reprinted with permission from (Kong et al., 2020) under Creative Commons License. Copyright © 2020 Kong et al.)

so as to ensure the scaffold withstands its major applied load (Kong et al., 2020). Alternative strategies for increasing the stiffness of the scaffold by increasing the density of the polymer and reducing the porosity simply increase the van de Waals forces, which are relatively easy to break. Visser et al. found that increasing the fiber thickness, changing the polymer density, or reducing the porosity does not endlessly improve the stiffness of the composite (Visser et al., 2015). Thick fibers have relatively less surface-to-volume ratio compared to thin fibers and thus reduce the interstitial connection with the surrounding gel. Boere et al. found that by covalently modifying the PCL matrix with a methacrylate group, one can significantly improve the interstitial bonding between the fiber matrix and GelMA hydrogel, which in turn improved the scaffold's resistance to repeated axial and rotational forces (Boere et al., 2014).

Chapter 4
Tissue-Specific Applications of Biotextile Collagen Scaffolds

The application of biomaterials in tissue engineering is primarily as a scaffold that provides physical support and transmits signals to cells so they can regenerate tissue. An ideal scaffold should provide a microenvironment suitable for cell growth and tissue formation and sufficient mechanical properties and act as a reservoir for growth factors chemokines and cytokines, as well as being able to be replaced by cells in order to fully restore the tissue architecture (Law et al., 2017).

The scaffold design is an essential component in the tissue regeneration process. The interactions between the cells, the ECM, and the scaffold regulate cell phenotype and cell activities such as migration, proliferation, differentiation, gene expression, and protein secretion, and thus the design and fabrication of the scaffold design needs to be carefully planned (Law et al., 2017). The clinical needs and requirements for the repair of different types of tissue can vary considerably. Sometimes fibrous structures are preferred for load-bearing purposes, whereas sometimes a viscous gel is required to fill in the pores and/or void space after tissue loss. In nature, the composition, conformation, arrangement, and functions of ECM proteins in different tissues also vary. In load-bearing tissues such as tendons and ligaments, collagen bundles are built up hierarchically from collagen molecules, fibrils, fibril bundles, to crimped fascicles (McCullen et al., 2010). The ECM shows a triphasic response to tensile stress with toe, linear, and yield regions in the stress-strain curve (McCullen et al., 2010). In the meniscus, the alignment of collagen fibers is significantly different from tendons and ligaments. The collagen bundles in the innermost region of the meniscus are oriented in a radial pattern, while those in the surrounding region are aligned in a circumferential direction, reflecting its unique mechanical function (McCullen et al., 2010) (Table 4.1).

In this chapter, we summarize and analyze recent advances in tissue engineering applications. We identify current clinical problems, particularly for each type of tissue, and explain the specific tissue structure, function, and features. Based on different physiological and anatomical characteristics, we explain how engineering

F. Zhang et al., *Collagen-Based Biotextiles for Tissue Engineering*,
https://doi.org/10.1007/978-3-031-71563-1_4

Table 4.1 Pros and cons of collagen as a biomaterial in tissue engineering

Pros	Cons
Biocompatible	High cost of purification
Abundant and easily available	Batch of batch variation
Biodegradable (degradation can be regulated by cross-links)	Risk of disease transmission (e.g., bovine spongiform encephalopathy)
Malleable into various forms (e.g., sheets, tubes, sponges, foam, nanofibers)	Variability in enzymatic degradation rate (depending on enzyme type and concentration)
Synergic with bioreactive components	Difficult to maintain its dimension in vivo due to swelling
Compatible with synthetic polymers	Poor mechanical strength in vivo. Not suitable for load-bearing tissues, e.g., bone
Promotes blood coagulation	
Nonantigenic	
Nontoxic	
Easily modifiable	
Bioresorbable	

principles are applied to each specific type of tissue and how critical problems in scaffold design and tissue regeneration can be addressed. We also summarize the advantages of using collagen-based scaffolds, especially collagen-based biotextiles for each clinical application.

4.1 Skin Tissue Engineering

As the largest organ of the body, the primary function of the skin is to serve as a protective barrier against the environmental invasion of foreign substances and harmful pathogens from the external environment (Meyer, 2019). Specifically, it contains three layers, namely, the epidermis, the dermis, and the subcutis, which is also known as the hypodermis or superficial fascia (Meyer, 2019). The epidermis contains epithelial cells and is composed of keratinized, stratified squamous epithelium which contains no blood vessels. On the other hand, the dermis does contain blood vessels, nerves, and other structures, such as hair follicles and sweat glands, which are made of connective tissue produced by fibroblasts. The connective tissue is composed of an interconnected mesh of elastin and collagenous fibrous tissue. In particular, the skin contains roughly 80–90% type I collagen, 8–12% type III collagen, and less than 5% type V collagen, which is a key component for collagen fibrillogenesis. Abundant amounts of type III collagen have been found in the papillary dermis, which provides nutrients to the epidermis and helps control the temperature of the skin. Type III collagen is also the predominant form of collagen in the reticular dermis which forms the thick bottom layer of the skin (Menon et al., 2012).

Skin wounds are common injuries which can be caused by many different reasons, such as physicochemical or thermal damage and metabolic diseases such as diabetes. These pathological conditions result in dysfunction of the barrier role against the external environment, increase of the level of pain, serious disability, or even death. It has been reported that approximately 5.7 million people in the USA suffer wound lesions and the treatment cost is estimated to be 50 billion dollars annually (Araujo et al., 2020).

As mentioned previously, collagen is a major component of proteins which form the skin and other tissues of the human body. In addition to being the main component of skin tissue, it has an essential role in the control of the inflammatory response to injury and contributes to subsequent repair by stimulating cellular mitogenesis, differentiation, and migration (Rangaraj et al., 2011). Collagen is also involved in the interaction between enzymes which remodel the ECM, including matrix metalloproteinases (MMPs) and their tissue inhibitors (TIMPs) (Rangaraj et al., 2011). Besides, collagen plays a key role in maintaining the biological and structural integrity of the ECM, which makes it a perfect candidate for wound healing. Purified collagen can be sourced from bovine, porcine, or human tissues and can be formulated in several different ways, such as a sponge, hydrogel, fiber, or powder (Chattopadhyay & Raines, 2014). Table 4.2 lists the commercial forms of reconstituted collagen. Many studies have demonstrated that collagen has the ability to accelerate the rate of healing (Chattopadhyay & Raines, 2014). As a result, collagen-based wound dressings have been fabricated into a range of different structures, including hydrogels, electrospun fibrous webs, and 3D-printed scaffolds and have long been used to treat decubitus ulcers and burn wounds (Duan et al., 2018). Table 4.3 summarizes recent studies for wound healing using collagen components. They demonstrate that pronounced healing efficiency is improved with the usage of collagen as a carrier material.

Table 4.2 Commercial forms of reconstituted collagen

Collagen form	Name (company)
Partially purified skin	Gelfoam (Pfizer)
Collagen sponge	Helistat (Integra LifeSciences) Instat (Johnson & Johnson) ActiFoam (MedChem) SkinTemp (BioCor)
Collagen fiber	Helitene (Integra LifeScience) Instant Fibrillar (Johnson & Johnson) Avitent (Medichem)
Collagen power	BioCore (Medichem)
Collagen composite dressing	Fibracol (Johnson & Johnson) Biobrane (UDL Laboratories)
Hydrolyzed collagen	Chronicure (Derma Sciences)

Table 4.3 Recent studies of collagen-based scaffold for wound healing

Materials	Scaffold fabrication technique	Cell sources/ components	Major results	References
Type I collagen	Electrospinning type I from calfskin collagen cross-linked with glutaraldehyde vapor		High porosity, excellent mechanical strength, and high surface area-to-volume ratio; obtained favorable performance for cell attachment, growth, and proliferation	Rho et al. (2006)
Chitosan/ collagen sponge	Sponges of carboxymethyl chitosan grafted with collagen peptide by covalent coupling and freeze-drying		Sponge demonstrated pronounced wound healing efficiency, and treatment enhanced cell migration and promoted skin regeneration, thereby highlighting the potential application of these sponges for the treatment of burn injuries	Cheng et al. (2019)
Collagen film	The film was prepared by freeze-drying of the solution containing hGH and collagen	Human growth hormone (hGH)	Using collagen as a carrier material showed its excellent biocompatibility and mild conditions for preparation and can maintain sustained releasing of unstable proteins	Maeda et al. (2001)
Collagen matrices	Two commercial collagen matrices were used with cell culture of gene transfection fibroblasts	Fibroblasts	Histological analysis of matrices explained on Day 31 indicated the successful survival of genetically modified cells and ingrowth of blood vessels	Rosenthal and Köhler (1997)
Neurotensin (NT)-loaded collagen matrices	Collagen matrices were prepared by freeze-drying, followed by neurotensin coating	Neurotensin (NT)	NT-loaded collagen dressing promoted the speed of healing for control/ diabetic wounds, enhanced reepithelialization, and reduced inflammatory cytokine expression and inflammatory edema	Moura et al. (2014)
Electrospun tilapia skin collagen nanofibers	The collagen sponge was obtained from tilapia skin and electrospun into nanofibrous	Human keratinocytes	The raw material proved not to induce a severe immune response, and electrospun fibers promoted the adhesion, differentiation, and proliferation of human keratinocytes	Zhou et al. (2016)

(continued)

Table 4.3 (continued)

Materials	Scaffold fabrication technique	Cell sources/ components	Major results	References
Type I collagen with silver nanoparticles	Fabricated by blending a web of plain collagen nanofibers with AgNP-loaded collagen nanofibers using electrospinning	Silver nanoparticles	With the addition of AgNPs, the healing efficiency has been improved. Accelerated reepithelization, collagen production, and better wound contraction were observed	Rath et al. (2016)

Collagen is an important structural protein maintaining both biological and structural integrity for the ECM. Owing to its special properties, such as low immunogenicity and its ability to be molded into durable and biocompatible scaffolds, collagen plays a leading role in wound care. In addition, collagen has been developed as a protein system for targeted delivery of controlled release bioactive molecules to the wound site. This has been demonstrated to expedite progress in the field of wound healing and tissue regeneration (Chattopadhyay & Raines, 2014).

4.2 Corneal Tissue Engineering

The cornea is located on the outer surface in front of the eye. It protects the iris and lens and shields the eye from germs and dust and helps focus and refract light onto the retina (Kong et al., 2020). Corneal injury is the second most common cause of blindness. At the present time, corneal transplants are the most effective treatment to corneal injury. In 2013, more than 70,000 corneal transplants were performed in the USA, up 5.9% from 2012 (Eye Bank Association of America, 2013). Limited recourse of cadaveric donor tissue no longer meets the increased demand, and the design, fabrication, and clinical insertion of artificial corneas (keratoprostheses) continue to face many problems such as retro prosthetic membrane formation, which involves the formation of a membrane behind the prosthesis that obstructs the visual axis and leads to decreased vision. Other complications can involve chronic inflammation, infection, and even melting at the corneal prosthesis interface (Zhang et al., 2016). Therefore, fabricating a viable corneal equivalent using tissue engineering has attracted much attention and has made some significant progress.

Corneal tissue has a complex anatomy. It is transparent and lacks blood vessels, making it a challenging tissue to engineer. The cornea contains three layers, namely, the corneal epithelium, the stroma, and the endothelium. In between these layers, there is the Bowman's membrane that is a thick structure which prevents scratches from reaching the stroma, and there is the Descemet's membrane that maintains the corneal structure, homeostasis, transparency, and hydration and anchors the cornea to the endothelium (Ghezzi et al., 2015). The cornea contains abundant collagen matrices, making collagen an attractive material for corneal tissue engineering. In addition, there is a high concentration of leucine-rich proteoglycan that supports the interfibrillar spacing that contributes to corneal transparency and maintains corneal hydration (Ghezzi et al., 2015). A corneal equivalent should have equivalent light transmittance and structure to the native cornea and support normal visual function. The material and design should minimize the inflammatory response and support cell growth. It should also show sufficient tensile strength and elastic modulus to enable surgical handling. Table 4.4 summarizes the recent advances in the design and structure of collagen-based corneal equivalents.

Collagen has been widely studied for corneal tissue engineering given its ubiquity in native cornea. Yet, pure collagen gel does not present adequate mechanical strength and long-term durability (Aghaei-Ghareh-Bolagh et al., 2019; Hong et al., 2018). Cross-linking is one of the most widely used methods to strengthen the collagenous corneal equivalent and prolong its presence and function in the eye. Zhang et al. fabricated scaffolds for corneal tissue engineering by using type I rat tail collagen which was cross-linked in one of three ways, namely, with ultraviolet light and riboflavin, with genipin, or with EDC. The transmittance and thickness of the scaffold was comparable with the human cornea. The mechanical strength was improved significantly after cross-linking, which extended beyond that of the anterior stroma. The transmittance of visible light was also improved due to cross-linking, making it equivalent to the human cornea. However, cross-linking decreased transmittance of UV light, which was similar to an aged cornea (Zhang et al., 2016). Kishore et al. also found cross-linking using EDC and NHS significantly increased the mechanical properties of the corneal equivalent, but cross-linking did not adversely affect keratocyte viability, morphology, and proliferation (Kishore et al., 2016).

Blending other biomechanically durable materials with collagen is another approach to strengthen the resulting corneal equivalent. Instead of using pure collagen, collagen has been combined with various types of material to leverage the advantage of different components such as silk, tobramycin, gelatin, chitosan, and decellularized ECM (Goodarzi et al., 2019; Liu et al., 2014; Rafat et al., 2008; Hong et al., 2018). Long et al. mixed collagen with silk fibroin and cross-linked it with EDC/NHS to strengthen the collagen-based membrane as a corneal implant. The membrane provided optimal mechanical properties with high suture retention strength when loaded with 10 wt% silk fibroin, and after implantation, the structure promoted complete epithelialization in 35 ± 5 days without symptoms of rejection (Long et al., 2015).

Recreating the complex structure of the cornea has been one of the promising outcomes achieved by corneal tissue engineering. The corneal stroma contains about 90% of the corneal tissue, which is composed of orthogonally aligned collagen nanofiber lamella (Kong et al., 2020). These collagen fibers are uniform, parallel, tightly packed, and connected to one another by glycosaminoglycans (GAG) and proteoglycans (PGs) (Kong et al., 2020). This specific alignment of collagen fibers contributes to the transparency of the stroma and the regular functionality of the cornea. On the other hand, if the collagen fibers are randomly oriented, they will

Table 4.4 Recent advances of collagen-based corneal tissue regeneration

Materials	Scaffold fabrication technique	Cell sources	Major results	Status	References
Collagen	Electrochemical compacted hydrogel, cross-linked by EDC-NHS and genipin	Keratocytes	Electrochemical compaction was able to generate an aligned collagen fiber matrix. However, uncross-linked material degrades within 6 h with the presence of collagenase. Cross-linking significantly improved the strength and modulus of the material and prolonged the degradation duration	In vitro	Kishore et al. (2016)
Porcine corneal matrix	Decellularized ECM	Limbal epithelial cell-like cells and corneal endothelial cell-like cells derived from human embryonic stem cells	Cells pre-seeded on the porcine corneal matrix accelerated the reepithelialization process and reduced vasculature development	In vivo	Zhang et al. (2017)
Collagen	Compressed collagen gel reinforced by decellularized corneal lenticule (dCL)	Limbal epithelial stem cells (LESCs)	Embedded dCL significantly improved the saturability and prolonged the degradation of the construct compared to compressed collagen gel. The construct was able to support the expansion and stratification of the LESCs in vitro. A rabbit model of LCSD demonstrated its efficacy as an LESC carrier with regard to saturability and durability as well as supporting corneal epithelial regeneration	In vitro and in vivo: rabbit	Hong et al. (2018)

(continued)

Table 4.4 (continued)

Materials	Scaffold fabrication technique	Cell sources	Major results	Status	References
Collagen	3D-printing	Keratocytes differentiated from human turbinate derived mesenchymal stem cells	Increasing shear force led to aligned collagen fibrils. The collagen fibrils remodeled along the printing direction and resulted in lattice structure assembling native cornea 4 weeks postimplantation. Shear-induced aligned construction also improved transparency and reduced haze formation compared to non-patterned construction	In vivo: rabbit	Kim et al. (2019)
Collagen vitrigel	Layer by layer, vitrification	Acellular	A dual-layered collagen vitrigel containing biomimetic sBM and sSL presented high optical transparency and low haze formation and vascularization. The incorporation of sBM facilitated epithelial cell migration and restoration of epithelial barrier function. Adding pSIS ECM microparticles in the sSL improved its mechanical properties, as well as suppressed pro-inflammatory gene expression	In vitro and in vivo: rabbit	Wang et al. (2020)
Porcine UBM particulate	Decellularized ECM	Acellular	UBM injection significantly reduced scar formation, vasculature development at limbus, and myofibroblast activation after partial removal of corneal epithelium, which was regulated by eosinophils and elevated IL-4 production	In vivo: athymic (nude) mice	Wang et al. (2021)

Abb: *EDC* 1-ethyl-3-(3-dimethylaminopropyl) carbodiimide, *NHS* N-hydroxysuccinimide, *dCL* decellularized corneal lenticule, *LESC* limbal epithelial stem cell, *LSCD* limbal stem cell deficiency, *ECM* extracellular matrix, *sBM* synthetic Bowman's membrane, *sSL* stromal layer, *pSIS* porcine small intestinal submucosa, *UBM* urinary bladder matrix compromise visual function. Therefore, when designing a corneal equivalent, mimicking this orthogonally aligned structure in the cornea has been an attractive goal and a challenging strategy.

Kong et al. fabricated a corneal equivalent using a gelatin methacryloyl hydrogel reinforced with orthogonally aligned fibers made from a resorbable polyethylene glycol/polycaprolactone (PECL) copolymer (Kong et al., 2020). The electrohydrodynamic process called near-field electrospinning (NFES) which is based on a stable cone-jet mode was used to generate precisely aligned microfibers so as to mimic the orthogonal alignment of collagen fibers in native corneal tissue. The grid shape alignment of the microfibers improved light transmittance compared to randomly aligned fibers. The orthogonal alignment of the microfibers was also implicated to enhance mechanical properties as well as inhibit differentiation of keratocytes into fibroblasts. The fiber spacing of 100 μm was found to be optimal to balance the mechanical properties and cellular response. In an in vivo rodent model, the fiber embedded hydrogel induced complete regeneration of stromal tissue and restored the normal morphology, despite the residue of slowly degrading PECL fibers. Furthermore, by adding supplementary ascorbic acid, insulin, and β-FGF to the serum-free media, the limbal stromal stem cells (LSSCs) were able to maintain their keratocyte phenotype, and the expression of collagen VI was enhanced (Kong et al., 2020).

In addition to the mechanical stimulus introduced by the corneal equivalent, the cross talk between stromal cells and immune cells performs essential functions in corneal regeneration (Wang et al., 2021). When activated keratocytes differentiate into fibroblasts and myofibroblasts, the randomly oriented collagen fibers form scar tissue that compromises light transmittance and may even cause loss of vision (Kong et al., 2020; Wang et al., 2021). Therefore, reducing the activation of myofibroblasts is critical in preventing scar formation. By selecting the biomaterial at the corneal injury site to be ECM, the formation of scar tissue was significantly reduced due to a decline in vasculature formation and myofibroblast activation. This resulted from significantly elevated IL-4 production by eosinophils at the wound site 1 week after injury in urinary bladder matrix (UBM) treated mice compared to phosphate buffered saline (PBS) treated mice. IL-4 is a pro-regenerative cytokine that is highly involved in the type-2 immune response. In the Gata1−/− mice that lack eosinophils, the benefit of injecting UBM treatment was completely ablated, highlighting the crucial role of eosinophils in UBM-reduced scar tissue formation (Wang et al., 2021).

4.3 Ligament/Tendon Tissue Engineering

Tendons and ligaments are connective tissues that are composed of a highly oriented cellular component and ECM (Rieu et al., 2017). Tendons join muscles to the bone, whereas ligaments join bones to the bone. Each year, there are over 400,000 shoulder tendon and ligament repairs, more than 300,000 tendon and ligament surgical procedures repairing foot and ankle injuries, and over 100,000 anterior cruciate ligament (ACL) repairs performed in the USA alone (Ratcliffe et al., 2015).

Tendons and ligaments have a three-dimensional hierarchical structure (Fig. 4.1), with the collagen fibers arranged in a highly aligned and longitudinal manner parallel to the mechanical axis. This leads to the high tensile strength of tendons and ligaments, despite the fact that each level of this collagen-rich hierarchy is interspersed with different amounts of noncollagenous extracellular matrix (Kastelic et al., 1978; Thorpe et al., 2013). This three-dimensional structure enables tendons and ligaments to tolerate and transmit large tensile forces between muscles and bones without damaging the macroscopic structure.

The collagen fibrils in tendons and ligaments vary in diameter, ranging from about 30 nm to approximately 300 nm depending on the stage of development. They are stabilized by intermolecular cross-links and bind to adjacent microfibrils (Barnard et al., 1987; Canty & Kadler, 2005). These intermolecular cross-links often appear as five collagen molecules linked together to form a pentafibril. While these intermolecular and intramolecular cross-links are essential in stabilizing the collagen fibrils, they also contribute to the high tensile strength of both tendons and ligaments (Canty & Kadler, 2002). The collagen molecules are oriented longitudinally in a well-characterized quarter-staggered manner, with a 40-nm gap between fibrils. This banded appearance of the collagen fibrils provides a periodicity repeating every 67 nm, which is known as the "D" spacing or "D" period (Thorpe et al., 2015).

There are many different types of collagen present in tendons and ligaments, and among these, type I collagen is the most abundant, making up to 90% of the total collagen content. Type III collagen is the other major type of collagen in tendons, comprising approximately 10% of the collagen content in a normal healthy tendon. Its primary role is thought to be responsible for the fibrillogenesis of collagen as

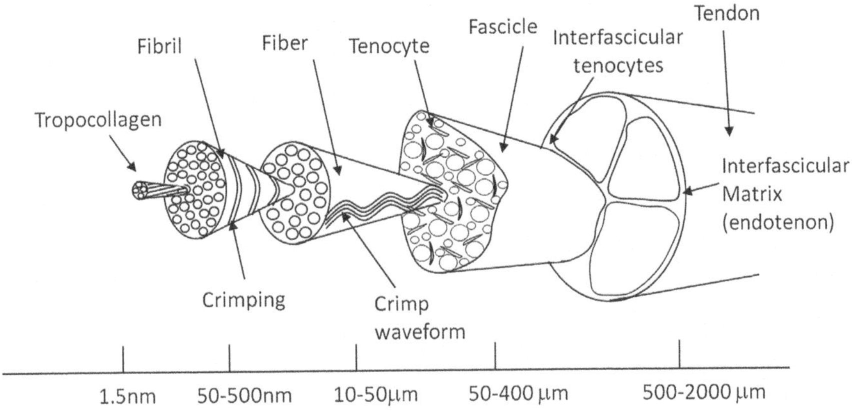

Fig. 4.1 Schematic diagram showing the hierarchical structure of a tendon. (Reproduced under Creative Commons Attribution from Harvey et al., 2009. Copyright © 2019 Harvey, Thompson, Cochlin, Raju, Cui, Cornell, Hulley & Brady)

well as regulating the size of the type I collagen fibrils (Thorpe et al., 2015). Both type I and type III collagen are fibrillar collagens made from three polypeptide chains. Type I collagen consists of two alpha1 chains and one alpha2 chain, expressed by two genes, specifically Col1A1 for the alpha1 chains and Col1A2 for the alpha2 chain. On the other hand, type III collagen has the gene Col3A1 expressed by three alpha1 chains (Adams, 1989). In addition to the collagens, there are some nonfibrous proteins found in tendons, such as the proteoglycans (PGs) which are the most abundant, making up 1–5% of the tendon dry weight. As the number of PGs increases in the tendon tissue, so also does the water content increase. This causes the tendon to become stiffer and to have increased resistance to compression (Berenson et al., 1996; Yoon & Halper, 2005) (Fig. 4.2).

When a tensile load is applied to tendons and ligaments, the crimped collagen fibers start to align parallel to each other, and they lose their crimped behavior, as observed in the toe region of the stress-strain curves. For tendons, the toe region is short (about 2–5%) due to their load transfer function, but for ligaments, which allow for different amounts of mobility in different joints, the toe region can vary from 4% to 40% depending on the anatomical site (Sensini & Cristofolini, 2018). If the load is applied directly to the collagen fibers, then the fibers will straighten and become oriented in the direction of the applied load. They will also exhibit elastic recovery when the load is removed, because they are operating in the linear region of the stress-strain curve. Beyond the linear region, the collagen fibers start to slide past each other which means that they lose the memory of their previously crimped structure and they become permanently deformed. At higher strains, individual collagen fibers experience progressive failure, until complete rupture of the tendon or ligament occurs (Wang, 2006; Woo et al., 1997, 2006).

There are various reasons and etiologies for tendon tears. They are mainly related to overuse as well as degenerative changes in tendon structure. Due to their location and use, certain tendons are more prone to failure than others. For example, the rotator cuff tendon, the Achilles tendon, the tibialis posterior and patellar tendons, and the exterior and flexor tendons in the hand and fingers are often subjected to direct laceration and tearing at all ages (Docheva et al., 2015). The situation for ligament injuries depends on the balance between stability and mobility of the joint. If there is excessive or repetitive transmission of force through the joint, leading to damage to other structures in and around the joint, then ligament injuries are likely to occur. Knees, hips, shoulders, ankles, elbows, and wrists are among some of the joints most commonly affected by ligament injuries (Hauser & Dolan, 2011).

Studies have shown that augmentation of a scaffold at the joint or failure site can reduce pain, increase the rate of healing, and lead to greater mobility and higher satisfaction and function compared to non-augmented tendon and ligament repairs (Hakimi et al., 2013). Considering the variability for different tendons and ligaments, scaffolds are expected to provide a microenvironment that mimics the native ECM environment for cell attachment, proliferation, and tissue regeneration (Chowdhury et al., 2018; Ratcliffe et al., 2015). More specifically, tissue engineered scaffolds need to fulfill the following requirements for tendon/ligament repair:

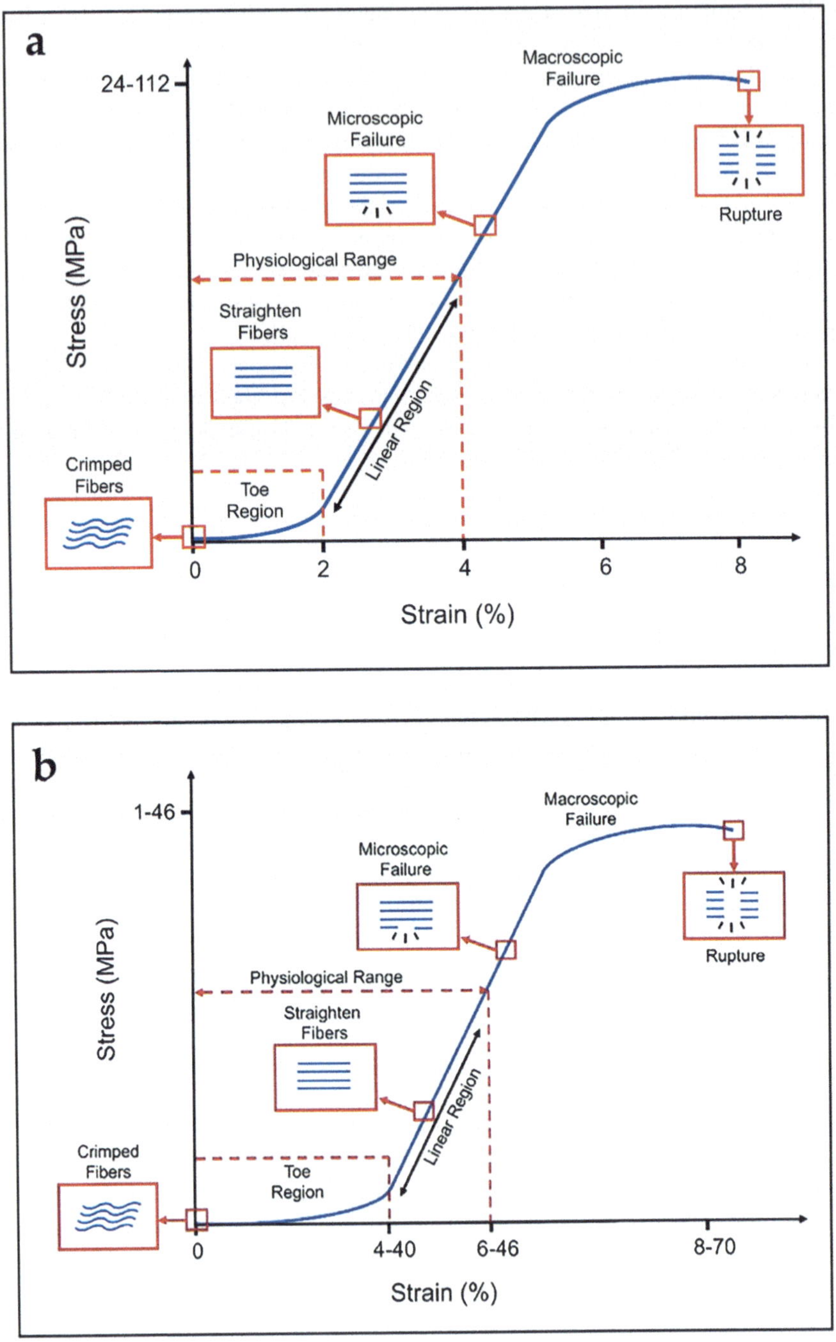

Fig. 4.2 Typical stress-strain curves that explain the behavior of collagen fibers in (**a**) tendons and (**b**) ligaments. (Reproduced under Creative Commons Attribution from (Sensini & Cristofolini, 2018). Copyright © 2018 Sensini & Cristofolini)

1. Biodegradability with an adjustable or tuned rate of degradation
2. Biocompatibility of the scaffold and its degradation products before, during, and after degradation
3. Superior mechanical properties that maintain their mechanical strength during the tissue regeneration process
4. Support for cell proliferation and differentiation, ECM secretion, and tissue formation
5. formation of desired constructs of complicated structures and shapes to meet the needs of the particular anatomical site (Liu et al., 2008) (Table 4.5)

As mentioned above, type I collagen is considered an ideal scaffold biomaterial due to being the major component of both tendons and ligaments. In order to mimic the conditions at the multiple tissue junction, it is important to design a scaffold which will promote the ECM to undergo continuous healing and remodeling. As seen in the above table for successful in vitro and in vivo studies, the collagen component needs to promote tendon cell attachment and proliferation, as well as migration and differentiation of stem cells. When the collagen fibers are well aligned, the biomechanical properties of the scaffold reach a clinically superior level compared to an unreinforced collagen sponge scaffold with inadequate mechanical performance (Zheng et al., 2017).

A freeze-drying technique is the most commonly used method for producing collagen scaffolds and sponges. The process involves freezing the collagen solution for several hours (usually overnight) at a low temperature of −60 to −80 °C and then freeze-drying under vacuum to produce a porous structure (Caliari et al., 2011; Chen et al., 2010). Weaving, knitting, and electrospinning have also been used to fabricate scaffolds for tendon and ligament repair (Islam et al., 2015; Maghdouri-White et al., 2018; Xie et al., 2021). And in order to enhance the mechanical properties of the scaffolds as well as to increase the resistance to enzymatic degradation, a physical dehydrothermal treatment has been proposed. It involves heating the collagen to a temperature in excess of 90°C while under vacuum to remove water. Alternatively, the use of 1-ethyl-3(3-dimethylaminopropyl) carbodiimide (EDC) and N-hydroxysuccinimide (NHS) chemical cross-linking technique has shown to stabilize the structure of collagen (Caliari et al., 2011; Chen et al., 2010; Islam et al., 2015; Xie et al., 2021).

Chen et al. incorporated human embryonic stem cells derived from mesenchymal stem cells (hESC-MSCs) within a knitted silk-collagen scaffold and applied them with mechanical stimulation in vitro. They found that the hESC-MSCs exhibited tenocyte-like morphology and expressed tendon-related gene markers. They also implanted the scaffolds in rats for Achilles tendon repair and observed the cells attached and proliferated along the collagen fibers under in vivo mechanical stimulus. In addition, the superior mechanical performance and the successful histological images demonstrated that hESC-MSCs seeded on silk-collagen scaffolds were able to promote tendon repair (Chen et al., 2010).

A nonstructural, highly porous collagen scaffold made from type I bovine collagen has been developed and has demonstrated that this scaffold can generate the

Table 4.5 Collagen-based scaffolds for tendon and ligament repair

Graft materials	Scaffold fabrication technique	Cell source	Development status	Tendon or ligament repair	Major results	References
Collagen and silk	Collagen: freezing method Silk: knitting	Rabbit bone marrow stem cells (BMSCs)	In vitro and in vivo (female New Zealand white rabbits, harvested at 8 and 12 weeks)	Rotator cuff tendon (supraspinatus tendon)	1. In vitro: rabbit BMSCs attached and spread well on collagen scaffolds 2. In vivo: collagen scaffolds combined with knitted silk had a positive effect on healing at the tendon and bone interface	Qian et al. (2019)
Collagen and silk	Electrospun	Human tenocytes and adipose-derived stem cells (ASCs) +BMP-13	In vitro		Silk fibroin-collagen scaffolds were attached Significantly more tenocytes compared to pure silk fibroin scaffolds; significantly higher cell migration through silk-collagen scaffolds with the presence of PDGF-BB and BMP-13	Maghdouri-White et al. (2018)
Collagen and silk	Knitting	Human embryonic stem cell-derived mesenchymal stem cells (hESC-MSCs)	In vitro and in situ (rat Achilles tendon)		hESC-MSCs seeded on knitted silk-collagen scaffolds differentiate into the tenocyte-lineage; dynamic mechanical stimulation is beneficial to tissue-engineered tendons	Chen et al. (2010)
Collagen and glycosaminoglycan (GAG)	Lyophilization	Tendon cells (TC)	In vitro		Tendon cells showed greater levels of attachment, metabolic activity, and alignment as well as less cell-mediated scaffold contraction on anisotropic collagen-GAG scaffolds	Caliari and Harley (2011)

Graft materials	Scaffold fabrication technique	Cell source	Development status	Tendon or ligament repair	Major results	References
Collagen	Weaving		In vivo (female New Zealand white rabbit)	Rotator cuff tendon	The load-bearing feature of the scaffold repair may help to improve the recovery of mechanical strength and create the potential for the resident cells to receive mechanical stimulation	Islam et al. (2015)
Collagen and silk	Knitting	Rabbit tendon stem progenitor cells (TSPCs)	Female rabbit (harvested at 4, 8,12 weeks)	Rotator cuff tendon	Aligned supracellular structure enhanced cellular infiltration and tenogenic differentiation and also provided a conductive healing environment for endogenous cell migration and differentiation	Zheng et al. (2017)

formation of new tendon tissue consisting of well-organized collagen fibers in a sheep model (Van Kampen et al., 2013). Research scientists have followed 23 patients who received a surgical procedure to repair their large or massive rotator cuff tears using an inductive collagen patch. The reports indicate that the patch did not cause any complications, and it promoted the formation of new tendon tissue as confirmed by ultrasound and magnetic resonance imaging (MRI). Although there were two incidences of clinical failure, and one of the patients underwent additional surgery, the overall healing rate at the 2-year follow-up was 96% (Thon et al., 2019).

4.4 Cartilage Tissue Engineering

Damage and degeneration to the knee and hip cartilage contributes significantly to disability and a lack of mobility all over the world (Kwon et al., 2019). Cartilage, including articular cartilage and the meniscus, lacks vasculature and a perfuse blood supply. As a result, the ability to repair and regenerate cartilage on its own post-injury is extremely limited (Shen et al., 2021). There is also a high risk that cartilage will experience a degenerative disease such as osteoarthritis after long-term wear, tear, and injury (Kwon et al., 2019).

As shown in Fig. 4.3, cartilage possesses unique structure and properties. Articular cartilage is a load-bearing tissue located at the end of long bones and contains a large amount of water (up to 70%), along with 20% of type II collagen and 10% of proteoglycans and low density chondrocytes (Kwon et al., 2019). Cartilage is inhomogeneous, anisotropic, multiphase, and viscoelastic and enriched by collagen fibers that reinforce its mechanical performance. It is composed of three regions, all of which present distinct fiber architecture, proteoglycan composition, and chondrocyte phenotype and density, corresponding to their distinct mechanical requirements (Liao et al., 2013; McCullen et al., 2010). The superficial zone, which is the surface of articular cartilage, is composed of collagen fibers that are aligned parallel to the surface of the subchondral bone. This fiber alignment provides the structure with high tensile modulus (~10MPa). On the other hand, the low proteoglycan content makes the superficial zone compliant in compression. All these features allow the superficial zone of the cartilage to withstand shear and tension caused by applied physiological loading (McCullen et al., 2010). In the middle zone underneath the superficial zone, the collagen fibers are oriented randomly, while in the deep zone, they are again aligned perpendicular to the subchondral bone. Compared to the superficial zone, the middle and deep zones show significantly lower tensile moduli of 3.2 MPa and 0.87 MPa, respectively (McCullen et al., 2010).

The meniscus is another type of cartilage tissue that is susceptible to damage. It is composed of an interlacing network of type I collagen fibers (Wang et al., 2020). The bulk of the meniscus can be divided into two distinct regions. In the innermost region, the collagen bundles are patterned in radial directions, whereas elsewhere the collagen fibers are oriented circumferentially (McCullen et al., 2010). Only a third of the meniscus in the peripheral area is vascularized, whereas other areas lack

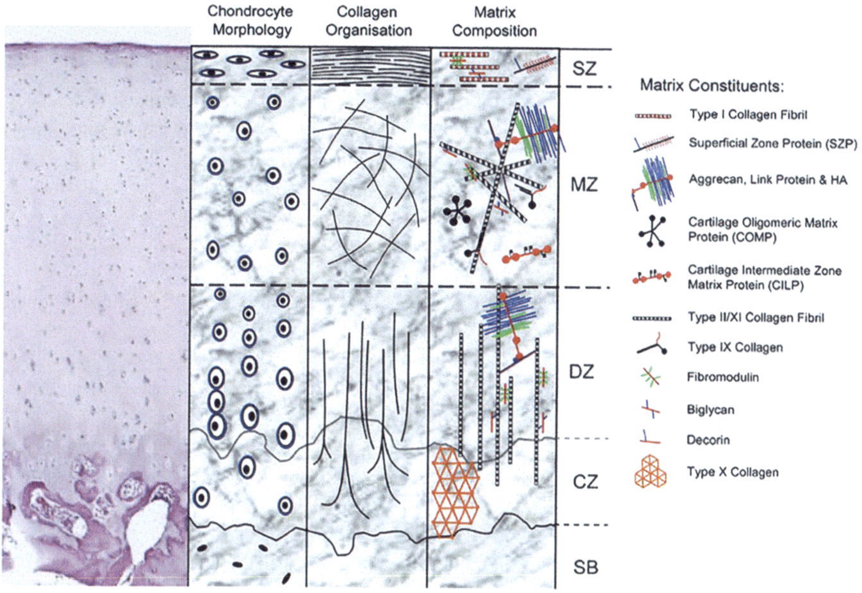

Fig. 4.3 Histological staining of hyaline cartilage and a schematic diagram of chondrocyte morphology and ECM structure in different regions of cartilage. Abbreviations: *SZ* superficial zone, *MZ* middle zone, *DZ* deep zone, *CZ* calcified zone, *SB* subchondral bone. (Reprinted under Creative Commons Attribution License from Francis et al., 2018. Copyright © 2018 Francis, Di Bella, Wallace and Choong)

a continuous blood supply, making it almost impossible for meniscal tears to self-repair after injury (Pabbruwe et al., 2010).

Current surgical treatments aim to repair small defects in articular cartilage using bone marrow stimulation, grafting, and cell therapy (Kwon et al., 2019). However, these therapies are not able to provide a long-term solution. Tissue engineering is considered as the most promising technique to completely heal and regenerate articular cartilage and damaged menisci after injury. Many materials have been utilized to fabricate tissue engineering scaffolds for cartilage repair, including both synthetic polymeric materials and natural materials. Using a tissue engineering scaffold provides immediate mechanical support as well as a cell and growth factor delivery carrier to promote tissue regeneration. In this section, we discuss the recent focus on cartilage tissue engineering and summarize the use of collagen-based materials as a scaffold material to promote articular cartilage and meniscal repair. Table 4.6 summarizes recent studies on collagen-based biomaterials for cartilage tissue engineering.

In order to improve the tissue engineering outcome, one needs to use the scaffold to provide cells with local environmental cues and instructions. Mimicking the three-dimensional ECM architecture is one of the most important approaches to provide cells with a native tissue-like microenvironment. An ideal scaffold would have high collagen density, aligned collagen fibrils, and anisotropic mechanical

Table 4.6 Recent advances in collagen-based cartilage tissue regeneration

Materials	Scaffold fabrication technique	Cell source	Major results	Status	References
Bovine menisci ECM	Decellularized and blended with polyethylene glycol diacrylate	Human bone marrow stem cells (hBMSCs)	The ECM derived from the inner meniscus enhanced the fibrocartilaginous differentiation of hBMSCs, characterized by the enhanced expression of type II collagen. ECM derived from the outer meniscus facilitated fibroblastic phenotype, characterized by increased expression of type I collagen by hBMSCs	In vitro	Shimomura et al. (2016)
Collagen	Biaxial plastic compression	Adipose-derived mesenchymal stem cells (MSCs)	The compressed collagen gel was densified and highly aligned. The highly aligned collagen fiber resulted in anisotropy of the modulus, which resembled many connective tissues. Using a hydrophilic elastomeric porogen, high porosity was achieved to allow cell infiltration, migration, and proliferation	In vitro	Zitnay et al. (2018)

(continued)

Table 4.6 (continued)

Materials	Scaffold fabrication technique	Cell source	Major results	Status	References
GelMA, alginate, polycaprolactone (PCL)	Interpenetrating hydrogel reinforced by 3D-printed PCL fiber network	Coculture of bone marrow-derived stromal cells and chondrocytes	The composite scaffold possessed equilibrium and dynamic mechanical properties that matched or approached that of native articular cartilage. It also supported the development of a hyaline cartilage-like tissue that was positive for GAG and collagen type II deposition while negative for type X collagen and calcium deposition	In vitro	Schipani et al. (2020)
GelMA, PCL, PEG	Hydrogel reinforced by melt electrowriting (MEW) Triblock polycaprolactone-polyethylene glycol copolymer (PCEC)	MSCs	A trilayer stratified scaffold was fabricated. PCEC fiber reinforcement significantly increased the mechanical strength compared to GelMA hydrogel. The stratified fiber alignment and localized growth factor delivery induced native tissue-like distribution of cell phenotype and matrix deposition in vitro. Cartilage and subchondral bone regeneration were achieved in the scaffold at the same time in vivo	In vitro and in vivo: rabbit	Qiao et al. (2021)

(continued)

Table 4.6 (continued)

Materials	Scaffold fabrication technique	Cell source	Major results	Status	References
High-density collagen gel	Collagen fiber deposition by in vitro cultured cells under mechanical constraints using clamped constructs	Rat meniscal fibro-chondrocytes	The native sized collagen fiber with a diameter of 20–40 μm was generated with hierarchical alignment by meniscal fibro-chondrocytes in vitro. The collagen fibers were organized in the same way that assembled juvenile native collagen from the fibril to fiber level after 6-week in vitro cell culture. These fibers possessed native fibril banding patterns and hierarchical fiber bundles of 50–350 μm in diameter. Its tensile properties improved significantly with time, reaching ~1 MPa	In vitro	Puetzer et al. (2021)

Abbreviations: *ECM* extracellular matrix, *MSCs* mesenchymal stem cells, *hBMSCs* human bone marrow mesenchymal stem cells, *GAG* glycosaminoglycans, *GelMA* gelatin methacryloyl, *PCL* poly(ε-caprolactone), *PEG* poly(ethylene glycol), *PCEC* triblock copolymer of ε-caprolactone and ethylene glycol

behavior that resembles native cartilage (Zitnay et al., 2018). Three-dimensional collagen sponges lyophilized and cross-linked from a 0.5% collagen solution provided embedded cells with a highly porous 3D structure and increased the expression of cartilage-specific genes. Compared to a 2D monolayer culture, the 3D collagen sponge was more successful in inducing differentiation of chondrocytes toward a specialized fibroblastic phenotype which was maintained over time (Glowacki & Mizuno, 2008). The highly porous collagen scaffold was also found to facilitate the delivery and migration of human adult bone marrow mesenchymal stem cells for meniscal repair (Pabbruwe et al., 2010).

The mechanical properties of the scaffold, such as stiffness, also influence the mechano-transduction of the embedded cells and further impact the regeneration outcome. Visser et al. found that a fiber-reinforced hydrogel scaffold improved the stiffness of the scaffold to a similar level to articular cartilage ranging from 400 to 800 kPa. Compared to the non-reinforced hydrogel, the reinforced scaffold upregulated the gene expression of embedded chondrocytes related to matrix deposition, including ACAN for glycosaminoglycan synthesis and COL1A1 for type I collagen (Henkel et al., 2013; Visser et al., 2015).

4.5 Skeletal Muscle Tissue Engineering

The skeletal muscle is the main tissue in animals, because it enables movement of the many and various body parts. It is the most abundant tissue type in the human body, comprising 60% of the total body weight (Cittadella Vigodarzere & Mantero, 2014). Apart from the skeletal muscle in the limbs and abdomen, the facial skeletal muscle has a unique purpose, where multiple functions such as chewing, speaking, and facial expressions are dependent on the movement of the facial skeletal muscles (Relaix & Zammit, 2012). Thus, the skeletal muscle is the most active and functional tissue structure (Cittadella Vigodarzere & Mantero, 2014).

Multiple reasons for skeletal muscle loss include but are not limited to genetic diseases such as cancer, accidents, severe burns, and trauma. Reconstructive procedures are required for treatment of congenital and orthopedic disorders and aging (Duan et al., 2021; Liu et al., 2018).

The skeletal muscle has its own supply of muscle precursor cells called satellite cells, inside the basal lamina of muscle fibers (Somers et al., 2017). These satellite cells are dormant, and in case of a small injury, these cells are activated and utilized for rapid regeneration of the lost tissue (Somers et al., 2017). But, in case of larger injury such as volumetric muscle loss due to disease or trauma, the regenerative process of satellite cells cannot work (Somers et al., 2017). Thus, researchers have been studying different ways of engineering a tissue construct that can be utilized for regeneration of large muscle tissues in vitro to be implanted as grafts in repairing volumetric muscle loss. Even if there is no FDA approved device currently available, many studies are dedicated to finding out the best possible scaffold biomaterial for in vitro facial skeletal muscle regeneration and implantation as a graft. Various synthetic and natural polymers have been extensively studied for this application including collagen, which is the leading natural biodegradable polymer being researched for craniofacial as well as skeletal muscle tissue regeneration among other natural polymers such as bioactive ceramics, glass, and hydrogels (Borrelli et al., 2020).

Collagen is the main protein found in the native muscle tissue; hence, it is found to be a promising candidate for tissue engineering in skeletal muscle regeneration (Somers et al., 2017). Collagen in its various forms is found to be effective in enhancing attachment and proliferation of skeletal muscle cells and inducing their differentiation in different types of cell culture systems. Cimetta et al. fabricated a perfusion bioreactor system using 3D collagen sponges to obtain a culture condition that can maintain long-term viability of large constructs of the skeletal muscle (Cimetta et al., 2007). They cultured C2C12 and skeletal muscle precursor cells individually in this system and obtained positive results in terms of cell distribution throughout the scaffold structure and long-term viability (Cimetta et al., 2007).

To replicate the structure and function of a skeletal muscle tissue, a tissue engineering scaffold is required to have a highly organized structure capable of providing the mechanical properties required for regeneration of functional skeletal muscle tissue. Textile structures incorporating bioresorbable polymers can thus provide an effective scaffold structure with anisotropic properties and controllable physical attributes including high total porosity, the desired pore size distribution, and biologically active surface characteristics. There have been multiple studies, where polymeric structures were incorporated into scaffolds for the regeneration of skeletal muscle tissue for the treatment of different kinds of muscle loss (Tables 4.7 and 4.8).

Five different types of natural polymeric materials including type I collagen, agarose, fibrin, alginate, and chitosan were studied by Pollot et al. (2018) for their physical and mechanical properties and ability to support myogenic activity. They fabricated hydrogels using the above 5 polymers and cultured primary satellite cells that were isolated from calves and the posterior legs of wild male Lewis rats. They evaluated the mechanical performance, such as the elastic modulus, and the in vitro myogenic properties of the individual hydrogels. They found that out of the five natural polymers, collagen and fibrin provided the most favorable structural support for long-term viability of muscle cells and showed favorable mechanical performance of the scaffold in terms of 100% elasticity and optimum elastic modulus (Pollot et al., 2018).

Shah et al. (2016) fabricated a composite scaffold using parallel aligned glass fibers coated with rat tail collagen. They evaluated the biocompatibility of the scaffolds in terms of cell alignment and proliferation using a mixed population of muscle-specific and non-muscle-specific human masseter muscle derived cells in a 21-day cell culture. The results indicated that due to the parallel aligned glass fibers, the cells were highly aligned unidirectionally. And the improved ability of the scaffolds to regenerate skeletal muscle was seen by skeletal muscle-specific gene upregulation (Shah et al., 2016).

Along with the scaffold structure which mimics the natural environment of the tissue, it has also been found that some amount of electrical and mechanical stimuli can provide the required biological cues for regeneration of a functional skeletal muscle tissue construct. To achieve these results, Sirivisoot et al. (2014) developed conductive gel-type scaffolds by incorporating polyaniline (PANI) and poly(3,4-ethylenedioxythiophene) (PEDOT) conductive nanofibers into their collagen solutions, together with the cell suspension. The resulting nanofiber-structured scaffolds demonstrated positive results in terms of biocompatibility with PC-12 (rat nerve cells) and hSMC (human smooth muscle cells) as well as proliferation and viability during the 7-day cell culture period. Interestingly, they also reported that the

Table 4.7 Collagen-based scaffolds for skeletal muscle regeneration

Scaffold materials	Scaffold fabrication technique	Cell sources	Status	Major results	References
Collagen coated glass fibers	Custom method used for coating rat tail collagen onto parallel aligned glass fibers	Mixed population of human masseter muscle cells containing myogenic and non-myogenic cell population	In vitro	Achieved unidirectional cell alignment with higher gene upregulation with collagen coated scaffolds	Shah et al. (2016)
Collagen sponge-porcine skin Type I and type III collagen	Freeze-drying	No cells	In vivo: rabbit model	No regeneration of larger skeletal muscle, but controlled bleeding, reduced inflammation and excessive scar tissue formation in 24 weeks	Kin et al. (2007)
3D collagen composite scaffolds	Obtained from BD Biosciences	C2C12 myoblasts (ATCC)	In vitro and in vivo (mice model)	3D skeletal muscle constructs were prepared in vitro, then improved muscle healing observed and vascularization with slow scaffold degradation in vivo	Ma et al. (2011)
3D microgrooved porous collagen	Liquid dispensing and freeze-drying, micropatterning with concave microgrooves	Rat L6 skeletal muscle myoblasts (ATCC)	In vitro	Controlled cell seeding and microgroove size and shape resulted in highly aligned and multilayered muscle bundle tissue formation	Chen et al. (2015)
Methacrylic acid (MAA) based collagen hydrogels	30% MAA and 70% collagen ratio by weight and carbodiimide cross-linking		In vivo (mouse model)	Enhanced muscle regeneration, greater muscle innervation	Carleton et al. (2021)

(continued)

Table 4.7 (continued)

Scaffold materials	Scaffold fabrication technique	Cell sources	Status	Major results	References
Poly caprolactone (PCL) collagen electrospun scaffolds	Electrospinning—1:1 by weight blend ratio of PCL and type I collagen (calf skin) solution	Human skeletal muscle cells (hSkMCs)	In vitro	Unidirectional tissue alignment and enhanced muscle fiber formation	Choi et al. (2008)

Table 4.8 Commercially available collagen scaffolds used for skeletal muscle engineering

Name of device	Type of collagen	Brief description of product	References
XenMatrix™ Surgical Graft (C. R. Bard, Inc.)	Porcine dermal collagen	Acellular non-cross-linked porcine dermal matrix	Dziki et al. (2016)
Biodesign® Otologic Repair Graft (Cook Medical)	Types I, III, VI collagen from porcine source	Laminated extracellular collagen matrix derived from porcine small intestinal submucosa	Dziki et al. (2016)
Integra® Dermal Regeneration Template (Integra Life Sciences)	Type I atelocollagen bovine tendon collagen	Porous matrix of thick collagen fibers covered with silicon layer	Frame and Frame (2006)
Integra® Meshed Dermal Regeneration Template (Integra Life Sciences)	Type I atelocollagen bovine tendon collagen	Bilayer matrix made of dermal layer of collagen and glycosaminoglycan and epidermal layer of pre-meshed polysiloxane	
Omnigraft Dermal Regeneration (Integra Life Sciences)		Inner porous, three-dimensional matrix, comprised of bovine collagen and chondroitin-6-sulfate (C6S) and thin polysiloxane outer layer	
Collaplug®, absorbable collagen wound dressing for dental surgery		Absorbable collagen material used as a hemostatic agent and wound dressing in oral surgery	

addition of the nanofibers into the collagen gel constructs improved the cell proliferation by more than four times compared to collagen gel constructs without polymeric nanofibers (Sirivisoot et al., 2014).

In yet another study reported by Choi et al. (2008), the researchers examined the feasibility of PCL/collagen electrospun scaffolds for implantable skeletal muscle constructs. The scaffolds were electrospun from a blended solution of PCL and collagen, using a 1:1 blend ratio by weight. The authors reported that they could obtain electrospun meshes oriented unidirectionally, which enhanced the alignment and proliferation of human skeletal muscle cells (hSkMCs) (Choi et al., 2008).

Collagen is the main protein found in the ECM of native tissue. It is thus the most important material to be incorporated into scaffold constructs for skeletal muscle regeneration. Moreover, as suggested above, having a polymeric backbone incorporated in the scaffold design improves alignment of the cells and supports the otherwise fragile collagenous structure with mechanical stability. Thus, blending the collagen material in any form with polymeric and textile structures has been found to provide a highly oriented and mechanically stable structural foundation for support and long-term viability of the tissue construct and improve the chances of regenerating functional skeletal muscle.

4.6 Muscle-Tendon Junction Tissue Engineering

While the literature reports major successes when tissue engineering a single cell type or tissue, at a multiple tissue junction, the tissue engineering strategy is more challenging because of its complexity. At a junction between two tissues, each tissue has very specific mechanical and biological functions. The junction tissue regeneration is much more than two neighboring tissues side-by-side, and it requires specific cellular structures and unique gene expression. For example, at a muscle-tendon junction (MTJ), the interface is required to transfer the force from the muscle to the tendon and only regenerates new tissue under sufficient mechanical stimulation.

He (2016) developed a 3D warp knitted spacer fabric with multiple sections to mimic the different elasticities, initial moduli, and tensile strengths that represented the muscle and tendon tissues at the two ends of the construct. Due to the high-tension requirements in warp knitting process, the materials were conventional PLA and PET yarns. The scaffold was cocultured with C2C12 skeletal myoblasts and 3T3 fibroblasts in a dynamic bioreactor under cyclic stretch and recovery loading. The cocultured scaffold detected the specific gene expression markers paxillin and vinculin, which confirms the successful regeneration of both MTJ tissues (He, 2016). Koeck fabricated a 3D woven scaffold out of collagen fibers extruded from type I and type III collagen. After coculturing with myoblasts and fibroblasts, the cells were observed to proliferate and penetrate the 3D structure. The strong and porous woven structure allowed the myofibers to align and stretch and recover. The results showed contractile activity of myotubes which indicated the successful formation of myofibers with striated morphology representing the natural skeletal muscle fibers. Both cells fused on the woven scaffold and formed an integrated interface structure (Koeck et al., 2022).

4.7 Cardiac Tissue Engineering

The heart wall is composed of tightly packed myocytes and fibroblasts, with dense supporting vasculature and collagen-based ECM.

The ECM is a protein complex secreted by fibroblasts that supports cell-cell adhesion, cell signaling communication, and other cell activities. The ECM is composed mainly of structural proteins such as type I and III collagen and elastin. Collagen is the predominant component of ECM, playing an important role in maintaining attachment between smooth muscle cells and providing the contractile function and tensile strength of the myocardium (Alberts et al., 2002). Myocardium consumes large amounts of oxygen and is hypoxia intolerant due to the high density and high metabolic demand of the cells (Vunjak-Novakovic et al., 2010). A heart attack, also known as a myocardial infarction (MI), happens when there is an insufficient oxygen supply to one or more areas of the heart muscle due to the blockage of blood flow in the coronary arteries. And because the self-renewal capacity of adult cardiomyocytes is limited, the loss of myocardium is irreversible, and the function and structure of the heart is hard to restore after a myocardial infarction (MI) (Park et al., 2005; Serpooshan et al., 2013).

Heart disease and stroke, the principal components of cardiovascular disease, account for nearly 40% of all deaths, more than all cancers combined (Vunjak-Novakovic et al., 2010). More than 3 million people suffer from MI every year in the USA, and about 5% of the population over 75 years old have had an MI with little or no history of symptoms. Left ventricle reconstruction (LVR), surgical ventricular restoration (SVR), and left ventricular assist device (LVAD) surgery are the most common surgical treatments for patients with end-stage heart failure (Athanasuleas et al., 2004; Klotz et al., 2008). The former technique is to remove the highly fibrotic tissue and implant a permanent patch to build the left ventricle, while the latter reverses the remodeling process and improves the global pump function by an LVAD pump, which is powered by an external power supply. Over half the patients have an LVAD implant procedure while waiting for a heart transplant, and while the cost of this procedure can reach as high as $200,000 in the first year, recurrence of the MI may occur within that time period (Rogers et al., 2012). With the greater understanding of stem cells and tissue engineering, stem cell therapy has become an attractive option for reversing left ventricular remodeling with a low cost, less invasive method. Currently, there are three ways to deliver cells to the wall of the left ventricle: (1) injection or infusion of the cell suspension, (2) injection of the cell suspension in a hydrogel that polymerizes in situ, and (3) implantation of a cardiac tissue patch to serve as a cell delivery vehicle at the site of the MI.

Due to the low stem cell retention rate by peripheral or intramyocardial injection, it is crucial to deliver the stem cells to a location that mimics not only the structural properties but also biological environment of native myocardial ECM, to provide support for these cells' survival and differentiation and the establishment of functional connections with the host tissue. And due to the limited mechanical support that a hydrogel can provide, a tissue engineered patch is the preferred choice.

Cardiac tissue engineering aims to create tissue constructs that are thick and compact, containing physiologic densities of metabolically active cardiac cell populations (Vunjak-Novakovic et al., 2010). An ideal cardiac patch is expected to match not only the chemical, mechanical, and biological properties but also the extracellular matrix morphology of native tissue. In this way, the patch provides immediate and lasting support for cardiac cell regeneration after a myocardial infarction

(Prabhakaran et al., 2011). It is notable that the stiffness of infarcted myocardial tissue changes from soft to rigid with time due to the formation of collagenous scar tissue, and consequently, the stiffness of the infarct bed determines the responses of implanted or recruited cells (Zhang et al., 2004). Type I collagen is the most abundant protein among the various cardiac ECM constraints, and examples of recent collagen-based scaffolds are listed in the following table (Table 4.9).

Lyophilization was used as the fabrication method to produce collagen scaffolds in the early 2000s. Cortes-Morichetti et al. injected human umbilical cord blood mononuclear cells (HUCBCs) along with lyophilized collagen matrix into mice (Cortes-Morichetti et al., 2007). The matrix had a distribution of pores dimensions between 50 and 100 μm and achieved a homogeneous cell distribution in the collagen pores after cell seeding.

Xiang et al. mixed type I collagen with chondroitin 6-sulfate and degassed and lyophilized the mixture to form a highly porous, 2-mm-thick scaffold sheet which was cross-linked by dehydrothermal (DHT) treatment followed by carbodiimide cross-linking (Xiang et al., 2006). After fabrication of the collagen/glycosaminoglycan scaffold, mesenchymal stem cells (MSCs) were seeded on the scaffolds and implanted in the regions of the rat heart where infarcts were induced 3 weeks earlier. The results indicated that the scaffolds cross-linked with DHT and carbodiimide could maintain the spongelike architecture through the 3-week implantation period and promote neovascularization at the site of implantation.

Compressed collagen hydrogels, collagen sponges loaded with stem cells, and cardiac patches containing stem cell-derived progenitor cells have indicated positive healing outcomes in animal models (Gao et al., 2011). Serpooshan et al. applied static compressive stress on hydrated collagen gels to remove water and produce a dense biomaterial with improved biological and mechanical properties (Serpooshan et al., 2013). Electrospinning has been widely used as the fabrication method to produce patches for cardiac tissue engineering, which can provide nanofiber webs with fiber diameters in the micron range to mimic the native ECM structure (Pham et al., 2006).

Kitsara et al. studied the different electrospinning parameters of clinically approved collagen in order to produce the optimal nanofiber scaffolds for three-dimensional culture of cardiac cells (Kitsara et al., 2015). They found that atelocollagen can produce a more uniform and consistent quality of electrospun fibers than either acidic or basic fibrous collagen. By designing a collagen scaffold with a high cross-link concentration and a longer post-electrospinning thermal exposure time, the surface of the scaffold became stiffer. This provided a more favorable environment for the cardiac cells to attach and proliferate.

Similar work was performed by Joanne et al. They electrospun collagen scaffolds from atelocollagen and treated them with citric acid. Different concentrations of the cross-linking agent and different durations for post-electrospinning thermal treatments were applied to the collagen scaffolds. Similar results were found to Kitsara's previous study when 10% citric acid and 150 min of thermal treatment were found to be optimal for cardiac tissue engineering. The research also indicated that human induced pluripotent stem cell-derived cardiomyocytes (hiPS-CMs)

Table 4.9 Collagen-based scaffolds for cardiac tissue engineering

Scaffold materials	Scaffold fabrication technique	Cell sources	Status	Major results	References
Collagen and glycosaminoglycans (GAG)	Degassed and lyophilized	Mesenchymal stem cells (MSCs)	In vivo	(N-(3-Dimethylaminopropyl)-N'-ethyl carbodiimide hydrochloride. EDAC group appeared to be largely intact, while dehydrothermal (DHT) cross-linked collagen-GAG scaffolds degraded A substantial amount of neovascularization was seen in the infarcted region in the implant groups and in the scaffolds	Xiang et al. (2006)
Collagen	Lyophilized	Human umbilical cord blood mononuclear cells (HUCBCs)	In vivo	Fragments of the collagen matrix thickened and protected the infarct scars, and the segments were consistently aligned along the left ventricular wall The matrix was completely absorbed at Day 45 without leaving a significant residual scar	Cortes-Morichetti et al. (2007)
Collagen	Gel	–	In vivo	The engineered patch was grafted onto the infarcted left ventricular myocardium immediately post myocardial infarction and significantly improved ejection fraction, and fractional shortening was found 4 weeks after the MI; functional networks of newly formed blood vessels were also found in the patch. Cell migration and angiogenesis were observed without a significant immune response	Serpooshan et al. (2013)
Collagen	Electrospinning	H9C2 cell line	In vivo (mouse model)	Atelocollagen can produce better quality of electrospun fibers than acid and basic fibrous collagen The cell culture performance can be improved by adjusting the cross-linking conditions The biocompatibility of the scaffold was excellent, and no visible inflammation was observed up to 2 weeks after implantation	Kitsara et al. (2015)

Scaffold materials	Scaffold fabrication technique	Cell sources	Status	Major results	References
Collagen	Electrospinning	Human induced pluripotent stem cell-derived cardiomyocytes (hiPS-CMs)	In vivo	Collagen scaffold seeded with hiPS-CMs remained stable after being implanted in mice for 14 days showing increased endothelial cell migration and greater vascularity	Joanne et al. (2016)
Collagen, elastin, polycaprolactone (PCL)	Electrospinning	C-kit+ cells	In vivo	The 80% collagen and elastin hybrid electrospun sheet extracted from the nucleus pulposus (NP) possessed a high capability for inducing c-kit+ cell proliferation. C-kit+-seeded 80% NP/PCL promoting transplanted cells underwent cardiac differentiation after 4 weeks	Liu et al. (2016)
Collagen, carbon nanofibers (CNF)	Lyophilized	H9C2 cell line	In vivo	Fibrosis also disappeared, and CD31 marker showed an increase in angiogenesis in the ischemic region in groups that have received the collagen only scaffold The immunohistochemical analysis of the heart tissue for platelet-derived growth factor receptor alpha (PDGFRα) is the most specific marker for fibro-adipogenic progenitors (FAPs) which showed a significant number of heart cells, including germ cells derived from the epicardium formed in the scaffold group with both collagen and CNF	Tashakori-Miyanroudi et al. (2020)
Collagen, polylactic acid (PLA)	Knitting	Conventional dendritic cells (cDCs)	In vivo	A knitted collagen/PLA heart cap produced a Young's modulus close to rat tissue which gave better cDC attachment than for the pure PLA knitted samples The heart cap implant was able to maintain the delivery of the rat cDCs to the infarct site during a period of 3 weeks, which provided a positive effect on promoting left ventricular function and reducing the size of the infarct zone after MI	Chen (2020)

seeded on collagen scaffolds could be utilized for the stabilization of dilated cardiomyopathy (DCM) (Joanne et al., 2016).

However, one of the limitations of fabricating electrospun scaffolds from natural proteins is the need to cross-link with glutaraldehyde or other cross-linker to stabilize the structure of the scaffold. Some researchers have proposed electrospinning an ECM-based hybrid scaffold by combining natural proteins, including collagen, elastin, and gelatin, with a synthetic polymer, for example, poly(ε-caprolactone) (PCL). Liu et al. performed an additional study and optimized the ratio of natural proteins to PCL in order to make the scaffolds suitable for cardiac repair after myocardial infarction (Heydarkhan-Hagvall et al., 2008; Liu et al., 2016). Liu et al. found that using 80% natural proteins extracted from the nucleus pulposus (NP) compared to PCL was the most effective ratio regarding cytocompatibility and mechanical characteristics compared to the 20% NP and the 50% NP groups which did reduce the infarction area, but there were limited left ventricular remodeling and a higher Young's modulus. Also, cardiac c-kit+ progenitor cells seeded on 80% NP scaffolds induced the formation of a significantly thicker left ventricular wall and reduced collagen hyperplasia compared to the other NP scaffolds, indicating the potential of using these electrospun ECM-based hybrid scaffolds for cardiac tissue engineering.

Conductive carbon-based nanomaterials have been combined with collagen because their conductivity and mechanical properties make them a potential candidate for heart tissue regeneration (Tashakori-Miyanroudi et al., 2020). An Iranian research group fabricated scaffolds by lyophilization measuring 5 × 6 × 2 mm made from collagen and different concentrations of carbon nanofibers (Tashakori-Miyanroudi et al., 2020). Although cell viability decreased significantly at carbon nanofiber (CNF) concentrations of 0.2 and 0.4 mg/ml, no significant differences were found at concentrations of 0.8 and 1.6 mg/ml compared to the control. The two tests to evaluate the proteins encoded with this caspase3 gene and the terminal deoxynucleotidyl transferase-mediated dUTP nick end labeling (TUNEL) assay are both tests to confirm cell apoptosis. The use of a blended scaffold containing collagen and carbon nanofibers (CNF) proved less immunogenic than pure collagen. The presence of carbon nanofibers helped to improve the mechanical support to the damaged tissue. Increased angiogenesis and new heart muscle cells were also found in the blended scaffold group containing collagen and carbon nanofibers, indicating a possible new treatment for patients with cardiac ischemia.

Contrary to the lyophilization and electrospinning approach, Chen designed and knitted a heart cap from hybrid collagen and PLA yarns and was able to deliver rat cardiosphere-derived cells (CDCs) to the left ventricle in a rat MI model. The knitted heart cap not only provided sufficient mechanical support to the left ventricular wall but also promoted the attachment and proliferation of rat CDCs so as to accelerate the recovery of the rat heart function after MI (Chen, 2020).

4.8 Vascular Tissue Engineering

The circulatory system is composed of a network of arteries, veins, and capillaries that carries oxygenated blood to tissues and removes metabolic waste from tissues so as to maintain their normal function. Vascular disease leads to a malfunction of the arteries or veins that negatively affects blood circulation which can lead to local tissue failure and that causes severe disease. Two types of arterial disease, coronary artery disease (CAD) and peripheral artery disease (PAD), are caused by the buildup of lipid plaque against the interior vessel wall that leads to narrowing or occlusion of the artery. This will lead to a limited blood supply to cardiac tissue or peripheral limbs that can cause severe pain, stroke, and organ failure. To address this matter, tissue-engineered vascular grafts have been studied extensively to replace or bypass the diseased vessels.

But in order to restore a functioning vascular network, the scaffold must be able to withstand the required mechanical pressure at a natural vessel/synthetic scaffold anastomosis. And these requirements differ from those described above under the topic of tissue-engineered vascular graft.

In this section, both aspects of vascular tissue engineering will be covered. The clinical problem, design principles, and the state-of-the-art in vascular tissue engineering will be discussed and analyzed.

4.8.1 Tissue-Engineered Vascular Grafts (TEVGs)

Vascular stenosis results in limited blood supply to tissues and organs, leading to organ malfunction, tissue death, and organ failure. Creating a bypass or replacing the stenosed or occluded blood vessel is an important surgical intervention that can restore blood flow. Synthetic vascular prostheses available on the market are primarily made from expanded polytetrafluoroethylene (ePTFE), also known as Gore-Tex™, poly(ethylene terephthalate) (PET) known commercially as Dacron™, or polyurethane (PU). However, these materials do not integrate well with native vascular tissues, and as a result, they can lead to chronic inflammation and complications after implantation. The high stiffness of these materials results in mechanical mismatch between the prosthesis and native vessel which causes a stress concentration at the anastomotic site. Because of these shortcomings, current synthetic vascular grafts contribute to intimal hyperplasia and stenosis, which in turn reduces the patency rate of the graft and the long-term survival rate of the patient.

Due to the limited availability of autologous vessels for grafting and the poor viability of synthetic vascular prostheses, research on tissue-engineered vascular grafts (TEVGs) has expanded rapidly over the last two decades, and significant advances have been made. There are already several research groups that have proceeded to a clinical trial phase that presents the possibility of using TEVG for vascular replacement in peripheral arterial disease as well as for vascular access during hemodialysis (Niklason & Lawson, 2020; Dahl et al., 2011a, b; Zhang & King, 2022).

The aim of TEVG is to construct a graft that is structurally and biochemically similar to a native blood vessel that can maintain a high patency rate and growth potential. According to ISO 7198: 2016, a tubular vascular graft should be able to be anastomosed to a blood vessel, maintain adequate patency, provide material and structural integrity, ensure dimensional stability, and maintain hemostasis by preventing blood permeating through the wall. Over time, typically within 6–9 months, the scaffold slowly degrades or resorbs, the mechanical load is gradually transferred to the new tissue, and a viable independent vascular device is formed (Hibino et al., 2015).

Different from other tissues, such as the skin and cartilage, engineered vessels must provide adequate mechanical properties initially at the time of implantation and not rely on maturation over time. So upon implantation, the TEVG must readily withstand the primary mechanical load caused by the internal pulsatile blood pressure as well as in the axial direction at the anastomosis. Mechanical mismatch adversely affects graft patency and revascularization outcomes (Seifu et al., 2013). The difference in compliance between an engineered graft and a native vessel leads to flow turbulence, which promotes platelet activation, thrombogenesis, and intimal hyperplasia (Seifu et al., 2013). In addition, since the TEVG scaffold was designed to degrade or absorb over time, any change in mechanical properties is important and needs to be planned in advance. It must ensure enough strength and let the newly formed tissue gradually assume the mechanical load. According to Seifu et al., a TEVG should be able to withstand a bursting pressure of 1700 mmHg and resist fatigue by not showing significant dilation when exposed to cyclic physiological loading for at least 30 days in vitro or in vivo (Seifu et al., 2013). TEVGs have been designed for the replacement of different blood vessels including, but not limited to, the coronary, peripheral, and carotid arteries, as well as for arteriovenous access for renal dialysis patients. Different vessels require various properties due to their different location and function, which need to be considered in the design. For instance, the coronary arteries require the graft to have sufficient strength to withstand the pulsatile forces caused by the beating heart, to be flexible to accommodate attachment to the pulsating left ventricular wall, and to resist material fatigue with no significant dilation after millions of cardiac cycles.

The TEVG scaffold is designed to replace a deceased vessel and restore a continuous blood supply to downstream tissues and organs. At the same time, it provides a structural template for cells to attach, infiltrate, proliferate, and deposit ECM to form a new tissue structure that resembles a native vessel. The main reason for most graft failure is early restenosis. Initial mural thrombus formation, excessive neotissue growth otherwise known as intimal hyperplasia, and an acute inflammatory response are all related to restenosis. To facilitate fast, effective, and controlled revascularization, the scaffold design should be antithrombotic, elicit a moderate level of inflammation, encourage endothelial cell migration, and prevent excessive tissue regeneration so as to avoid intimal hyperplasia.

As seen in Table 4.10, a variety of different approaches have been investigated to fabricate viable TEVGs. In the table, we summarize recent studies using collagen as one of the components. Collagen is the most abundant protein in native vascular ECM and thus has excellent biocompatibility to vascular tissue and blood components, such as noncytotoxic, low antigenicity, and a low inflammatory response (Boccafoschi et al., 2005). It is primarily the responsibility of smooth muscle cells (SMCs) and fibroblasts to provide the mechanical support that blood vessels need so as to withstand the internal pulsatile pressure and prevent excessive dilation of the vessel over time (Boccafoschi et al., 2005). Collagen regulates homeostasis by maintaining blood vessel integrity. Excessive collagen deposition leads to vessel narrowing or even occlusion, such as during the plaque formation and the development of atherosclerosis in coronary artery disease. However, loss of collagen due to degradation can also lead to vessel rupture or aneurysmal formation (Manon-Jensen et al., 2016). Lu et al. coated ePTFE vascular graft with heparin and collagen using a layer-by-layer self-assembly technique. The incorporation of heparin and collagen significantly reduced the rate of hemolysis and platelet adhesion, prolonged plasma recalcification time, and improved endothelial cell adhesion compared to a bare ePTFE graft (Lu et al., 2013). We also coated a PLA vascular graft with a gelatin methacrylate hydrogel to improve endothelial cell migration and proliferation (Zhang et al., 2023b)

Aligned collagen nanofibers are known to affect the inflammatory response of endothelial cells (ECs). For example, a substrate containing 30-nm-diameter collagen fibers has been shown to downregulate the gene expression of intercellular adhesion molecule-1 (ICAM1) and monocyte chemoattractant protein-1 (MCP1), which leads to a reduction in monocyte adhesion (Huang et al., 2013). Huang et al. further engineered a bilayer vascular graft prototype with an inner layer lined by longitudinally aligned collagen nanofibers and an outer layer with aligned nanofibers wound circumferentially. The endothelial cells (ECs) seeded on the inner layer and the smooth muscle cells (MSCs) seeded on the outer layer both showed aligned morphology in the direction of the nanofibers. This result confirms that a tissue engineered vascular graft with the desired cell orientation and alignment can be achieved by designing and fabricating the appropriate fiber alignment (Huang et al., 2013).

In a murine peripheral artery disease (PAD) model, an EC-seeded collagen nanofiber scaffold was selected to improve blood perfusion and promote local neoangiogenesis near the scaffold (Nakayama et al., 2015). Integrin $\alpha 1$ was found to play an important role in improving endothelial cell (EC) outgrowth in this process, and by adhering to the collagen scaffold, more ECs were retained at the transplantation sites compared to the control sample which involved direct cell injection (Nakayama et al., 2015).

Although collagen has been found to benefit tissue regeneration outcomes, collagen is not the only protein in the ECM. A moderate level of collagen was suggested to be helpful, whereas excessive amounts could be harmful. A balanced composition would better mimic the ECM to improve tissue regeneration. Shojaee et al. electrospun collagen blended with PCL as a series of vascular grafts. It was

Table 4.10 Recent advances of TEVG containing collagen and its derivatives

Graft scaffold materials	Scaffold fabrication technique	Graft inner diameter (mm)	Cell sources	Development status	Major results	PI	References
Fibroblast deposited ECM	Cell-assembly extracellular matrix (CAM)	4.8	Autologous human fibroblast	Clinical trial (human)	Function as hemodialysis access for up to 20 months in five patients Primary patency was 78% after 1 month and 60% after 6 months postimplantation	L'Heureux N	McAllister et al. (2009)
Human cadaveric smooth muscle cell deposited ECM	Cell deposition and decellularization	6	Decellularized	In vivo (baboon)	Patent for up to 6 months Patency rate was 88% at 1 year postimplantation No dilation and change in thickness for 24 weeks No sign of infection, calcification, aneurysmal dilation, and substantial intimal hyperplasia	Niklason LE	Dahl et al. (2011a)
Allogeneic canine smooth muscle cell deposited ECM	Cell deposition and decellularization	3–4	Autologous endothelial cells (ECs)	In vivo (canine)	Patency rate was 83% at 1 year postimplantation No sign of infection, calcification, aneurysmal dilation, and intimal hyperplasia at anastomoses up to 1 year	Niklason LE	Dahl et al. (2011b)
Allogeneic canine smooth muscle cell deposited ECM	Cell deposition and decellularization	3–4	Autologous endothelial cells (ECs)	In vivo (canine left)	Patency rate was 83% Patent up to 1 month	Niklason LE	Dahl et al. (2011b)

Graft scaffold materials	Scaffold fabrication technique	Graft inner diameter (mm)	Cell sources	Development status	Major results	PI	References
Porcine small intestine submucosa (SIS)	Decellularized and shaped into a tube	–	Acellular	In vivo (canine)	Patent for up to 60 days in the carotid artery Explant vessels showed increased strength due to remodeling Burst pressures exceeded human vessels (up to 5600 mmHg)	Lantz	Sandusky et al. (1995) and Roeder et al. (2001)
Polycaprolactone (PCL) and collagen	Electrospun	–	Acellular	In vivo (lapine)	Patent for up to 1 month as rabbit aortoiliac bypass grafts Little cell invasion or thrombosis observed	Atala	Tillman et al. (2009)
Composite: polycaprolactone (PCL), polyurethane (PU), and collagen	Electrospun	–	Acellular	In vivo (canine)	Patent for up to 8 weeks in the canine femoral artery A thin layer of endothelial cells (ECs) formed in vivo	Zhang J	Lu et al. (2013)

(continued)

Table 4.10 (continued)

Graft scaffold materials	Scaffold fabrication technique	Graft inner diameter (mm)	Cell sources	Development status	Major results	PI	References
Polycaprolactone (PCL) and collagen	Electrospun preimplantation at rat peritoneal cavity	1.6	Acellular	In vivo (rat)	Patent for up to 6 weeks in rat abdominal aorta Increased collagen content increased intimal hyperplasia in vivo Increased collagen content decreased cell infiltration in vivo Preimplantation reduced intimal hyperplasia	Bashur CA	Shojaee et al. (2017)
Collagen	Electrochemical alignment thread, knitting, and electrospun	4–5	Human umbilical vein endothelial cells (HUVECs)	In vitro	Bursting strength exceeds vein grafts Compliance matches vein graft Support HUVEC adhesion and proliferation	King MW NC State University	Zhang et al. (2019)
Collagen and elastin	Lyophilization and cross-linking	–	Human umbilical vein smooth muscle cells (SMCs)	In vitro	Scaffold strength improved by mechanical stimulation Stress-strain curve partially matched native vessels	Feijen J, University of Twente	Engbers-Buijtenhuijs et al. (2006)

Graft scaffold materials	Scaffold fabrication technique	Graft inner diameter (mm)	Cell sources	Development status	Major results	PI	References
Collagen	Gel	–	Porcine smooth muscle cells (SMCs) and endothelial cells (ECs)	In vitro	Cell proliferation and collagen remodeling observed over 7 days Very low burst pressures achieved (about 18 mmHg)	Mantovani D, Laval University	Boccafoschi et al. (2007)
Collagen	Gel	–	Murine aortic smooth muscle cells (SMCs)	In vitro	Scaffold strength improved by increased collagen deposition as a result of mechanical stimulation Burst pressures remained well below those of natural blood vessels	Nerem RM Georgia Institute of Technology,	Schutte et al. (2010)
Chitosan and gelatin	Knitting and coating; lyophilization	–	Murine vascular smooth muscle cells (SMCs)	In vitro	Burst pressures of 4000 mmHg achieved Suture retention strengths exceeded that of the ovine carotid artery	Zhang X, Tsinghua University	Schutte (2010)

found that about 10% collagen was able to develop moderate levels of neointima while higher blend levels, around 25%, caused intimal hyperplasia compared to the no collagen control (Shojaee et al., 2017). The graft with the lower collagen content recruited less inflammatory macrophages and expressed less ECM proteins, including collagen and elastin, compared to the higher blend level of collagen and the no collagen control.

There are other proteins in the ECM that play pivotal roles in tissue engineering, such as elastin, which contributes to the contractility of the graft and prevents aneurysmal formation (Montero et al., 2020). To mimic the ECM composition for a better tissue engineering outcome, the use of decellularized ECM has also been extensively explored. Decellularized ECM is obtained by removing the native cellular components containing the DNA protein from the harvested vessels or tissue to eliminate any immune response after implantation. It retains a delicate balance of native ECM composition and structure, and therefore it can provide infiltrated cells with a microenvironment resembling native blood vessels. Carotid arteries of the African black ostrich are employed to create a decellularized vessel because of their straight geometry with a lack of branches. The luminal surface has been modified with an $\alpha 4\beta 1$ integrin ligand REDV tetrapeptide, which is present in the sequence of the type III connecting segment of human plasma fibronectin and is known to achieve long patency and neointimal formation (Montero et al., 2020; Shojaee et al., 2018). Porcine descending abdominal aortas have likewise been surface-modified by oriented heparin, to mimic the orientation of heparin sulfate on the endothelial cell surface, which has been proven to reduce thrombogenicity (Mahara et al., 2015). A porcine urinary bladder matrix (UBM) has also been used as a scaffold for endovascular aneurysm repair (Dimitrievska et al., 2015). The UBM retains the basement membrane which is thought to prevent endothelial cell penetration through the scaffold and promote the formation of an endothelial monolayer (Callanan et al., 2014). The use of cell deposited ECM protein as a graft material is another method that Niklarson and L'Heureux research groups have been actively exploring, as described previously in Table 4.10.

Preimplantation of a vascular graft in the rat peritoneal cavity was found to lead to cellular recruitment and remodeling of the graft. Therefore, after implantation in the rat abdominal aorta, reduced intimal hyperplasia was found in preimplanted grafts compared to bare control grafts (Callanan et al., 2014). Such implantation improvements were related to the early recruited cells and deposited ECM protein in the peritoneal cavity. Preimplantation also resulted in recruitment of inflammatory macrophages. Preimplantation allows scaffold to recruit cells as well as undertake remodeling in an in vivo bioreactor prior to implantation as a vascular graft. Cells in preimplanted grafts were found to express low levels of smooth muscle alpha-2 actin and ECM protein including collagen and elastin, which was in agreement with the results of reduced intimal hyperplasia. Decreased accumulations of oxidized lipid species were also observed in the preimplanted grafts (Shojaee et al., 2017). However, although all these factors appear to be associated with the reduction of intimal hyperplasia, a cause-effect relationship has yet to be identified.

4.8.2 Microvasculature Engineering

The microvasculature involves small vessels or capillaries that are developed deep inside and around the tissue and transport oxygen rich blood to and remove metabolic products from the tissues and kidneys. An important issue dealing with the survival and integration between engineered tissue and native tissue, especially the thick tissue constructs, needs to be discussed. The diffusion limitation of oxygen and nutrients is 200 μm, and therefore the maximum distance between a cell to the adjacent blood vessel should be less than 200 μm (Rouwkema & Khademhosseini, 2016). Cell viability has been found to be correlated with the oxygen gradient across the thickness of a major tissue (Shojaee et al., 2017). Insufficient perfusion of the engineered tissue leads to the absence of nutrients and oxygen and accumulation of metabolic wastes, resulting in limited success of cell therapy and the application of engineered tissue.

In order to improve the perfusion rate of the engineered tissue, different approaches have been adopted to improve blood perfusion of the tissue-engineered construct after implantation, which includes in situ angiogenesis and the use of a prevascularized tissue construct.

In in situ angiogenesis, the scaffold relies on the angiogenic properties of the construct and seeded cells. This is of vital importance for the repair of tissue loss using a tissue engineering strategy. Without fast restoration of a blood supply after implantation, the engineered tissue will experience ischemia and hypoxia, leading to significant cell death and subsequent necrosis of the engineered tissue. Angiogenesis is the process by which vascular development occurs from an existing vascular network. First, endothelial cell tips start to sprout, followed by capillary formation and arteriogenesis, when mural cells in the vascular wall, including smooth muscle cells (SMCs) and pericytes, proliferate in the outer layers of the vessel wall and develop mature vascular functions such as dilation and contraction. It is not only the amount of microvasculature development that contributes to a successful outcome but also the organization, planning, and execution of vessel wall maturation (Rouwkema & Khademhosseini, 2016).

Including cells with angiogenic potential is one approach to improve continuous and effective blood perfusion. Stem cells, such as bone marrow-derived stem cells (BMSCs), adipose-derived stem cells (ASCs), and dental pulp mesenchymal stem cells (DPSCs), have all been shown to promote angiogenesis (Gorin et al., 2016). Compared with fibroblasts, which can reduce endothelial cell (EC) migration and induce a low level of EC self-assembly into a capillary, DPSCs were found to induce ECs to form extended capillary networks both in vitro and in vivo, as observed by fibroblast growth factor-2 (FGF-2) and hypoxia (Gorin et al., 2016). Such angiogenic potential has also been found by the paracrine effect of vascular endothelial growth factor (VEGF) and hepatocyte growth factor (HGF). Although HGF alone was not able to cause capillary formation by ECs, HGF showed a dose-dependent effect on how far a capillary will generate in the presence of VEGF (Gorin et al., 2016). The absence of VEGF in DPSCs completely inhibited capillary formation and the release of HGF (Gorin et al., 2016). In addition, other researchers have found that DPSCs can directly contribute to vascular maturation by depositing cells that create the basement membranes (Atlas et al., 2021).

Given that the angiogenic effect of cell therapy is associated with paracrine growth factors and/or chemokines, then the release of growth factors from scaffolds that promote angiogenesis involves a totally different strategy. VEGF is the most important growth factor involved in the conceptualization, promotion, and process of angiogenesis. VEGF induces the formation and extension of filopodia as well as the expression of DLL4 protein in the tip cells, and filopodia engages with those of a nearby tip cell to form a "bridge" and the formation of a new vessel. VEGF gradients control tip cell migration, whereas stalk cell proliferation is regulated by the concentration of VEGF. The leading tip cell responds to a VEGF gradient by migrating outward from the parent vessel to guide the shape and number of sprouts. On the other hand, VEGF also regulates endothelial stalk cell proliferation, which decides the sprout length (Rouwkema & Khademhosseini, 2016). Therefore, it can be appreciated why VEGF is most commonly used in the evaluation and regulation of angiogenesis.

Although the initial sprouting is of vital importance to initiate the angiogenesis process, the following maturation of the vessels is even more important for a successful perfusion outcome. Therefore, the release of a single cytokine is not always enough to regulate angiogenesis. In the presence of VEGF, HGF promotes the length of vessel development in a dose-dependent manner (Atlas et al., 2021; Gorin et al., 2016). Basic fibroblast growth factor (bFGF) was found to affect not only the functioning of endothelial cells (ECs) but also smooth muscle cells (SMCs). Platelet-derived growth factor (PDGF) primarily regulates cells in the vessel wall (including SMCs and pericytes) and promotes vascular maturation (Rouwkema & Khademhosseini, 2016). Angiopoietin 1 (Ang1) stabilizes the blood vessel, tightens the EC junctions, and increases the EC survival rate by improving EC-mural cell interaction (Rouwkema & Khademhosseini, 2016). Ang2 competes with Ang1 for the same cell receptor, but its function depends on the presence of VEGF. In the presence of VEGF, Ang2 facilitates EC migration and proliferation and promotes basal lamina remodeling. Depending on the timing of the release, Ang2 causes endothelial cell (EC) death and vessel regression (Rouwkema & Khademhosseini, 2016). So, it can be understood that the spatiotemporal control of the timing and duration of the release of these growth factors from the scaffold remains a challenge.

The extracellular matrix is one of the main factors that determines the angiogenic properties of the engineered tissue. Without appropriate ECM features, the engineered tissue and vasculature are not able to form the native-like construct (Nakayama et al., 2019; Quarta et al., 2017). Moreover, the physical properties of the substrate material are relatively easy to alter compared to locally controlling the release of growth factors. Nakayama et al. found that the nanofibrillar collagen scaffold assembled with parallel alignment significantly improved both myogenesis and angiogenesis in vitro and in vivo. The aligned scaffold was also found to improve the orientation, length, and striated immunostained for myosin heavy chain (MHC) expression patterns of the myotubes formed on the scaffold, as well as promoting the magnitude of the contractility and synchronicity of the construct. Cocultured ECs on the aligned scaffold produced a significantly higher level of angiogenin, VEGF-A, and insulin-like growth factor binding protein-3 (IGFBP-3), all of which play important roles in vasculature formation. Another relevant observation was that the tissue with the aligned scaffold was found to have a significantly greater number of perfused vessels. When the scaffold was fabricated from aligned collagen fibers, the vessels were more ordered and oriented parallel to the myotubes, compared to the scaffold containing random fibers (Nakayama et al., 2019).

The stiffness of the substrate has also been observed to affect the production of functional endothelial cell (EC) proteins. Santos et al. found that while the stiffness of the substrate did not affect the proliferation and gene expression of ECs, the higher stiffness (3 vs. 30 kPa) reduced the expression of VEGFR-2 while increasing the release of caveolin-1 and β-catenin in ECs (Santos et al., 2015). Within a different and lower stiffness range (150Pa vs. 4000Pa), a similar effect on VEGFR-2 was observed at both the transcriptional and expressional level (Mammoto et al., 2009). All three of these proteins (VEGFR-2, caveolin-1, and β-catenin) play important roles in angiogenesis. VEGFR-2 is the main receptor of VEGF-A that mediates angiogenesis. Caveolin-1 is responsible for caveolae formation, which regulates VEGFR-2 recycling from the cell membrane. And inactive β-catenin coordinates with VE-cadherin to transfer mechanical stimuli and stabilizes EC tight junctions (Santos et al., 2015). Having said that, no evidence was found in this study to indicate there was a direct impact of stiffness on the development of the vascular networks in vivo.

Compared to in situ angiogenesis, the strategy of using prevascularized tissue has been developed as an alternative and potentially attractive solution that is more rapid and controllable. Prevascularized tissue relies on the sprouting of ECs and self-assembly capillaries in vitro and at the anastomosis between the engineered vasculature and native vessel in vivo (Atlas et al., 2021). The in vitro process simulates the angiogenesis process in vivo in a simplified manner. Briefly, it involves the sprouting of endothelial tip cells and the proliferation of endothelial stalk cells (Atlas et al., 2021). However, ECs are not the only cell type involved in this process. Depending on the culturing conditions and the other cocultured cell type, the pre developed vascular network ranges from an immature cord-like network or a mature vascular network with well-developed lumen (Sun et al., 2016). It is well established that mural cells, including smooth muscle cells (SMCs) and pericytes, stabilize the vessels and regulate its permeability (Rouwkema & Khademhosseini, 2016). Koike et al. found that engineered vessels containing only HUVECs and the collagen/fibronectin gel hardly survived and were barely perfused after in vivo implantation. On the contrary, vessels developed by HUVECS together with 10T1/2 mesenchymal precursor cells were mostly perfused (Rouwkema & Khademhosseini, 2016). Pericytes contributed to basement membrane deposition, vascular maturation, vascular permeability, and integrity (Atlas et al., 2021). Alta et al. observed that DPSCs, also a type of mesenchymal cell, resembled the morphology of pericytes and deposited basement membranes that promote maturation of the vasculature (Atlas et al., 2021). The angiogenic property of DPSCs derived from stem cells from human exfoliated deciduous teeth (SHED) also improved capillary formation in vitro. SHED can differentiate into many cell types, including adipocytes, chondrocytes, endothelial cells, neurons, and odontoblasts. It was found that coculturing SHED with endothelial cells effectively promoted mature microvasculature in engineered constructs (Atlas et al., 2021). The prevascularized tissue showed improved perfusion rate and therefore the survival rate of the tissue-engineered construct after implanting subcutaneously in the mouse (Atlas et al., 2021). Further study in more clinically relevant animal models would be needed to validate this clinical relevance.

While early studies relied on the self-assembly of ECs and mural cells to revascularize the scaffold in vitro, it has been found more effective to pattern the cell distribution network within the scaffold. This can be achieved using various scaffold fabrication technologies such as molding, electrospinning, coating, microfluidic technology, and 3D-printing (Koike et al., 2004).

Chapter 5
Concluding Remarks and Future Perspectives of Collagen-Based Biotextiles for Tissue Engineering Applications

Collagen, as the most abundant protein in extracellular matrix (ECM), has long been utilized in a wide range of tissue engineering applications. It provides a biological motif for cells to recognize and adhere to. At the same time, the fibrous morphology and intermediate mechanical properties of collagen also provide cells with a microenvironment that resembles native ECM.

Although the application of collagen to certain clinical end uses has long been limited due to its intermediate mechanical properties and difficulty in restoring the natural morphology, new technologies have been developed to overcome these problems. By leveraging newly developed technologies, we can tailor fibrous collagen materials for various needs and applications. Biotextile fabrication methods have long been used to produce the first-generation medical products such as vascular grafts, hernia meshes, and wound dressings. Here, we have highlighted different fiber-forming technologies including electrospinning, electrochemical alignment, fiber extrusion, as well as fiber-reinforced composite preparation methods and surface modification techniques. By selecting different hybrid materials, we have combined their advantages for tissue engineering applications. Leveraging more recently developed fiber-forming technologies, we can combine collagen materials with their excellent biological properties, with advanced biotextile fabrication methods to generate a more flexible yet robust textile structure. With different textile structures, complicated yet interconnected architectures can be designed to fulfill various requirements for different tissue applications.

Although collagen is the predominant protein in ECM, there are also other proteins that signal cells. The synergistic effect of ECM proteins in regulating the cellular response and regenerating tissue is yet to be fully understood. As a result, mixing collagen with other ECM proteins is an important concept that has yet to be studied. Combining various textile technologies, such as electrospinning, fiber extrusion, and composite fabrication using different materials blended together, offers a powerful and synergistic tool to optimize the conditions for a range of different tissue engineering end uses.

F. Zhang et al., *Collagen-Based Biotextiles for Tissue Engineering*,
https://doi.org/10.1007/978-3-031-71563-1_5

Careful selection and design of both the material and the structure is important to the ultimate success of the engineered tissue. When designing a collagen-based tissue engineered scaffold, the characteristics of the target tissue must be taken into account. The aligned fibrous distribution of the native ECM, the mechanical requirements of the target tissue, and the unique microenvironment at the target site will ultimately affect the design of the scaffold and the strategy for its fabrication. For those applications that require superior mechanical properties, such as the repair of ligaments and tendons, the mechanical strength and durability of a textile structure can be leveraged to provide the required performance. Blending natural collagen with resorbable synthetic materials is another approach to improve the mechanical performance of the scaffold. If the polymeric material is available as a fiber, filament, or yarn, then blending it with collagen or other ECM proteins offers an easy experimental strategy. However, when considering the design of a tissue engineered vascular graft, the extent of the elasticity and compliance of the graft wall has to be carefully considered. For corneal tissue regeneration in the eye, the optical properties of the scaffold material pose a critical question. Therefore, the use of one material and one structure that can address all the requirements is not possible, and so careful selection and design is the key to success.

References

Adams, S. L. (1989). Collagen gene expression. *American Journal of Respiratory Cell and Molecular Biology, 1*(3), 161–168.

Aghaei-Ghareh-Bolagh, B., Guan, J., Wang, Y., Martin, A. D., Dawson, R., Mithieux, S. M., & Weiss, A. S. (2019). Optically robust, highly permeable and elastic protein films that support dual cornea cell types. *Biomaterials, 188*, 50–62.

Agrawal, A., Rahbar, N., & Calvert, P. D. (2013). Strong fiber-reinforced hydrogel. *Acta Biomaterialia, 9*(2), 5313–5318.

Ahn, J. I., Kuffova, L., Merrett, K., Mitra, D., Forrester, J. V., Li, F., & Griffith, M. (2013). Crosslinked collagen hydrogels as corneal implants: effects of sterically bulky vs. non-bulky carbodiimides as crosslinkers. *Acta biomaterialia, 9*(8), 7796–7805.

Akbari, M., Tamayol, A., Bagherifard, S., Serex, L., Mostafalu, P., Faramarzi, N., & Khademhosseini, A. (2016). Textile technologies and tissue engineering: A path toward organ weaving. *Advanced Healthcare Materials, 5*(7), 751–766.

Akentjew, T. L., Terraza, C., Suazo, C., Maksimcuka, J., Wilkens, C. A., Vargas, F., et al. (2019). Rapid fabrication of reinforced and cell-laden vascular grafts structurally inspired by human coronary arteries. *Nature Communications, 10*(1), 1–15.

Akkus, O., Kishore, V., Younesi, M., & Islam, A. (2018). *Electrochemical processing of materials, methods and production*. US: Google Patents.

Alberts, B., Johnson, A., Lewis, J., Raff, M., Roberts, K., & Walter, P. (2002). Cell junctions. In *Molecular biology of the cell* (4th ed.). Garland Science.

Annis, D., Bornat, A., Edwards, R. O., Higham, A., Loveday, B., & Wilson, J. (1978). An elastomeric vascular prosthesis. *Transactions – American Society for Artificial Internal Organs, 24*, 209–214.

Araujo, T. A. T., Almeida, M. C., Avanzi, I., Parisi, J., Simon Sales, A. F., Na, Y., & Renno, A. (2020). Collagen membranes for skin wound repair: A systematic review. *Journal of Biomaterials Applications, 36*(1), 95–112. https://doi.org/10.1177/0885328220980278

Athanasuleas, C. L., Buckberg, G. D., Stanley, A. W., Siler, W., Dor, V., Di Donato, M., et al. (2004). Surgical ventricular restoration in the treatment of congestive heart failure due to post-infarction ventricular dilation. *Journal of the American College of Cardiology, 44*(7), 1439–1445.

Atlas, Y., Gorin, C., Novais, A., Marchand, M. F., Chatzopoulou, E., Lesieur, J., et al. (2021). Microvascular maturation by mesenchymal stem cells in vitro improves blood perfusion in implanted tissue constructs. *Biomaterials, 268*, 120594.

Avila Rodríguez, M. I., Rodriguez Barroso, L. G., & Sánchez, M. L. (2018). Collagen: A review on its sources and potential cosmetic applications. *Journal of Cosmetic Dermatology, 17*(1), 20–26.

F. Zhang et al., *Collagen-Based Biotextiles for Tissue Engineering*, https://doi.org/10.1007/978-3-031-71563-1

Bächinger, H. P., Mizuno, K., Vranka, J. A., & Boudko, S. P. (2010). 5.16 – Collagen formation and structure. In H. Liu & L. Mander (Eds.), *Comprehensive natural products II* (pp. 469–530). Elsevier.

Barnard, K., Light, N. D., Sims, T. J., & Bailey, A. J. (1987). Chemistry of the collagen cross-links. Origin and partial characterization of a putative mature cross-link of collagen. *Biochemical Journal, 244*(2), 303–309.

Beachley, V., & Wen, X. (2010). Polymer nanofibrous structures: Fabrication, biofunctionalization, and cell interactions. *Progress in Polymer Science, 35*(7), 868–892.

Bella, J., Eaton, M., Brodsky, B., & Berman, H. M. (1994). Crystal and molecular structure of a collagen-like peptide at 1.9 A resolution. *Science, 266*(5182), 75–81.

Berenson, M. C., Blevins, F. T., Plaas, A. H., & Vogel, K. G. (1996). Proteoglycans of human rotator cuff tendons. *Journal of Orthopaedic Research, 14*(4), 518–525.

Bi, L., Cao, Z., Hu, Y., Song, Y., Yu, L., Yang, B., et al. (2011). Effects of different cross-linking conditions on the properties of genipin-cross-linked chitosan/collagen scaffolds for cartilage tissue engineering. *Journal of Materials Science: Materials in Medicine, 22*(1), 51–62.

Birukova, A. A., Tian, X., Cokic, I., Beckham, Y., Gardel, M. L., & Birukov, K. G. (2013). Endothelial barrier disruption and recovery is controlled by substrate stiffness. *Microvascular Research, 87*, 50–57.

Boccafoschi, F., Habermehl, J., Vesentini, S., & Mantovani, D. (2005). Biological performances of collagen-based scaffolds for vascular tissue engineering. *Biomaterials, 26*(35), 7410–7417.

Boccafoschi, F., Rajan, N., Habermehl, J., & Mantovani, D. (2007). Preparation and characterization of a scaffold for vascular tissue engineering by direct-assembling of collagen and cells in a cylindrical geometry. *Macromolecular Bioscience, 7*(5), 719–726.

Boere, K. W., Visser, J., Seyednejad, H., Rahimian, S., Gawlitta, D., Van Steenbergen, M. J., et al. (2014). Covalent attachment of a three-dimensionally printed thermoplast to a gelatin hydrogel for mechanically enhanced cartilage constructs. *Acta Biomaterialia, 10*(6), 2602–2611.

Borrelli, M. R., Hu, M. S., Longaker, M. T., & Lorenz, H. P. (2020). Tissue engineering and regenerative medicine in craniofacial reconstruction and facial aesthetics. *The Journal of Craniofacial Surgery, 31*(1), 15–27.

Caliari, S. R., & Harley, B. A. (2011). The effect of anisotropic collagen-GAG scaffolds and growth factor supplementation on tendon cell recruitment, alignment, and metabolic activity. *Biomaterials, 32*(23), 5330–5340.

Callanan, A., Davis, N. F., McGloughlin, T. M., & Walsh, M. T. (2014). Development of a rotational cell-seeding system for tubularized extracellular matrix (ECM) scaffolds in vascular surgery. *Journal of Biomedical Materials Research Part B: Applied Biomaterials, 102*(4), 781–788.

Canty, E. G., & Kadler, K. E. (2002). Collagen fibril biosynthesis in tendon: A review and recent insights. *Comparative Biochemistry and Physiology Part A: Molecular & Integrative Physiology, 133*(4), 979–985.

Canty, E. G., & Kadler, K. E. (2005). Procollagen trafficking, processing and fibrillogenesis. *Journal of Cell Science, 118*(7), 1341–1353.

Carleton, M. M., Locke, M., & Sefton, M. V. (2021). Methacrylic acid-based hydrogels enhance skeletal muscle regeneration after volumetric muscle loss in mice. *Biomaterials, 275*, 120909.

Carr, B. P., Chen, Z., Chung, J. H., & Wallace, G. G. (2022). Collagen alignment via electro-compaction for biofabrication applications: A review. *Polymers, 14*(20), 4270.

Casali, D. M., Yost, M. J., & Matthews, M. A. (2018). Eliminating glutaraldehyde from cross-linked collagen films using supercritical CO2. *Journal of Biomedical Materials Research Part A, 106*(1), 86–94.

Cauich-Rodriguez, J. V., Deb, S., & Smith, R. (1996). Effect of cross-linking agents on the dynamic mechanical properties of hydrogel blends of poly (acrylic acid)-poly (vinyl alcohol-vinyl acetate). *Biomaterials, 17*(23), 2259–2264.

Cha, B. H., Shin, S. R., Leijten, J., Li, Y. C., Singh, S., Liu, J. C., et al. (2017). Integrinmediated interactions control macrophage polarization in 3D hydrogels. *Advanced Healthcare Materials, 6*(21), 1700289.

Chakrapani, V. Y., Gnanamani, A., Giridev, V. R., Madhusoothanan, M., & Sekaran, G. (2012). Electrospinning of type I collagen and PCL nanofibers using acetic acid. *Journal of Applied Polymer Science, 125*(4), 3221–3227.

Charulatha, V., & Rajaram, A. (2003). Influence of different crosslinking treatments on the physical properties of collagen membranes. *Biomaterials, 24*(5), 759–767.

Chattopadhyay, S., & Raines, R. T. (2014). Collagen-based biomaterials for wound healing. *Biopolymers, 101*(8), 821–833.

Chaudhuri, O., Gu, L., Klumpers, D., Darnell, M., Bencherif, S. A., Weaver, J. C., et al. (2016). Hydrogels with tunable stress relaxation regulate stem cell fate and activity. *Nature Materials, 15*(3), 326–334.

Chen, J. (2020). *Preclinical evaluation of a biodegradable knitted heart cap for the delivery of cardiosphere-derived cells to induce reverse remodeling of a failing left ventricle.*

Chen, J. L., Yin, Z., Shen, W. L., Chen, X., Heng, B. C., Zou, X. H., & Ouyang, H. W. (2010). Efficacy of hESC-MSCs in knitted silk-collagen scaffold for tendon tissue engineering and their roles. *Biomaterials, 31*(36), 9438–9451.

Chen, S., Nakamoto, T., Kawazoe, N., & Chen, G. (2015). Engineering multi-layered skeletal muscle tissue by using 3D microgrooved collagen scaffolds. *Biomaterials, 73*, 23–31.

Cheng, Z., & Teoh, S. (2004). Surface modification of ultra thin poly (ε-caprolactone) films using acrylic acid and collagen. *Biomaterials, 25*(11), 1991–2001.

Cheng, X., Gurkan, U. A., Dehen, C. J., Tate, M. P., Hillhouse, H. W., Simpson, G. J., & Akkus, O. (2008). An electrochemical fabrication process for the assembly of anisotropically oriented collagen bundles. *Biomaterials, 29*(22), 3278–3288.

Choi, J. S., Lee, S. J., Christ, G. J., Atala, A., & Yoo, J. J. (2008). The influence of electrospun aligned poly(ε-caprolactone)/collagen nanofiber meshes on the formation of self-aligned skeletal muscle myotubes. *Biomaterials, 29*(19), 2899–2906.

Chowdhury, S. R., Busra, M. F. M., Lokanathan, Y., Ng, M. H., Law, J. X., Cletus, U. C., & Idrus, R. B. H. (2018). Collagen type I: A versatile biomaterial. In *Novel biomaterials for regenerative medicine* (pp. 389–414). Springer.

Cimetta, E., Flaibani, M., Mella, M., Serena, E., Boldrin, L., De Coppi, P., & Elvassore, N. (2007). Enhancement of viability of muscle precursor cells on 3D scaffold in a perfusion bioreactor. *International Journal of Artificial Organs, 30*(5), 415–428.

Cittadella Vigodarzere, G., & Mantero, S. (2014). Skeletal muscle tissue engineering: Strategies for volumetric constructs. *Frontiers in Physiology, 5*, 362.

Colomer, I., Chamberlain, A. E. R., Haughey, M. B., & Donohoe, T. J. (2017). Hexafluoroisopropanol as a highly versatile solvent. *Nature Reviews Chemistry, 1*(11), 1–12.

Cortes-Morichetti, M., Frati, G., Schussler, O., Van Huyen, J. D., Lauret, E., Genovese, J. A., et al. (2007). Association between a cell-seeded collagen matrix and cellular cardiomyoplasty for myocardial support and regeneration. *Tissue Engineering, 13*(11), 2681–2687.

Dahl, S. L., Kypson, A. P., Lawson, J. H., Blum, J. L., Strader, J. T., Li, Y., et al. (2011a). Readily available tissue-engineered vascular grafts. *Science Translational Medicine, 3*(68), 68ra9.

Dahl, S. L., Blum, J. L., & Niklason, L. E. (2011b). Bioengineered vascular grafts: Can we make them off-the-shelf? *Trends in Cardiovascular Medicine, 21*(3), 83–89.

Deitzel, J. M., Kleinmeyer, J., Harris, D. E. A., & Tan, N. B. (2001). The effect of processing variables on the morphology of electrospun nanofibers and textiles. *Polymer, 42*(1), 261–272.

Delgado, L. M., Pandit, A., & Zeugolis, D. I. (2014). Influence of sterilisation methods on collagen-based devices stability and properties. *Expert Review of Medical Devices, 11*(3), 305–314.

Dewle, A., Pathak, N., Rakshasmare, P., & Srivastava, A. (2020). Multifarious fabrication approaches of producing aligned collagen scaffolds for tissue engineering applications. *ACS Biomaterials Science & Engineering, 6*(2), 779–797.

Dimitrievska, S., Cai, C., Weyers, A., Balestrini, J. L., Lin, T., Sundaram, S., Hatachi, G., Spiegel, D. A., Kyriakides, T. R., Miao, J., Li, G., Niklason, L. E., & Linhardt, R. J. (2015). Click-coated, heparinized, decellularized vascular grafts. *Acta Biomaterialia, 13*, 177–187.

Docheva, D., Müller, S. A., Majewski, M., & Evans, C. H. (2015). Biologics for tendon repair. *Advanced Drug Delivery Reviews, 84*, 222–239.

Dong, B., Arnoult, O., Smith, M. E., & Wnek, G. E. (2009). Electrospinning of collagen nanofiber scaffolds from benign solvents. *Macromolecular Rapid Communications, 30*(7), 539–542.

Dos Santos, L. V., Cruz, M. R., Lima Lopes, G. D., & Lima, J. P. D. S. N. (2015). VEGF-A levels in bevacizumabtreated breast cancer patients: A systematic review and meta-analysis. *Breast Cancer Research and Treatment, 151*, 481–489.

Duan, L., Yuan, Q., Xiang, H., Yang, X., Liu, L., & Li, J. (2018). Fabrication and characterization of a novel collagen-catechol hydrogel. *Journal of Biomaterials Applications, 32*(7), 862–870.

Duan, K., Gao, X., & Zhu, D. (2021). The clinical relevance and mechanism of skeletal muscle wasting. *Clinical Nutrition (Edinburgh, Scotland), 40*(1), 27–37.

Dziki, J., Badylak, S., Yabroudi, M., Sicari, B., Ambrosio, F., Stearns, K., et al. (2016). An acellular biologic scaffold treatment for volumetric muscle loss: results of a 13-patient cohort study. *NPJ Regenerative Medicine, 1*(1), 1–12.

Engbers-Buijtenhuijs, P., Buttafoco, L., Poot, A. A., Dijkstra, P. J., de Vos, R. A., Lotus, M., et al. (2006). Biological characterisation of vascular grafts cultured in a bioreactor. *Biomaterials, 27*(11), 2390–2397.

Engler, A. J., Sen, S., Sweeney, H. L., & Discher, D. E. (2006). Matrix elasticity directs stem cell lineage specification. *Cell, 126*(4), 677–689.

Engler, A. J., Carag-Krieger, C., Johnson, C. P., Raab, M., Tang, H., Speicher, D. W., et al. (2008). Embryonic cardiomyocytes beat best on a matrix with heart-like elasticity: Scar-like rigidity inhibits beating. *Journal of Cell Science, 121*(22), 3794–3802.

Eye Bank Association of America. (2013). Eye banking statistical report.

Feitosa, N. M., Richardson, R., Bloch, W., & Hammerschmidt, M. (2011). Chapter 8 – Basement membrane diseases in zebrafish. In H. W. Detrich, M. Westerfield, & L. I. Zon (Eds.), *Methods in cell biology* (pp. 191–222).

Fernández-Colino, A., & Jockenhoevel, S. (2020). Textile-reinforced scaffolds for vascular tissue engineering. In B. Walpoth, H. Bergmeister, G. Bowlin, D. Kong, J. Rotmans, & P. Zilla (Eds.), *Tissue-engineered vascular grafts. reference series in biomedical engineering* (pp. 1–25). Springer.

Fernández-Colino, A., Wolf, F., Rütten, S., Schmitz-Rode, T., Rodríguez-Cabello, J. C., Jockenhoevel, S., & Mela, P. (2019). Small caliber compliant vascular grafts based on elastin-like recombinamers for in situ tissue engineering. *Frontiers in Bioengineering and Biotechnology, 7*, 340.

Fernández-Pérez, J., & Ahearne, M. (2019). The impact of decellularization methods on extracellular matrix derived hydrogels. *Scientific Reports, 9*(1), 1–12.

Fong, H., Chun, I., & Reneker, D. H. (1999). Beaded nanofibers formed during electrospinning. *Polymer, 40*(16), 4585–4592.

Frame, J. D., & Frame, J. E. (2006). Modifying integra as a regeneration template in deep tissue planes. *Journal of Plastic, Reconstructive & Aesthetic Surgery, 59*(5), 460–464.

Francis, L., Giunco, F., Balakrishnan, A., & Marsano, E. (2010). Synthesis, characterization and mechanical properties of nylon–silver composite nanofibers prepared by electrospinning. *Current Applied Physics, 10*(4), 1005–1008.

Francis, S. L., Di Bella, C., Wallace, G. G., & Choong, P. F. (2018). Cartilage tissue engineering using stem cells and bioprinting technology—barriers to clinical translation. *Frontiers in Surgery, 5*, 70.

Gao, J., Liu, J., Gao, Y., Wang, C., Zhao, Y., Chen, B., et al. (2011). A myocardial patch made of collagen membranes loaded with collagen-binding human vascular endothelial growth factor accelerates healing of the injured rabbit heart. *Tissue Engineering Part A, 17*(21–22), 2739–2747.

Ghezzi, C. E., Rnjak-Kovacina, J., & Kaplan, D. L. (2015). Corneal tissue engineering: Recent advances and future perspectives. *Tissue Engineering Part B: Reviews, 21*(3), 278–287.

Gilbert, T. W., Sellaro, T. L., & Badylak, S. F. (2006). Decellularization of tissues and organs. *Biomaterials, 27*(19), 3675–3683.

Gilpin, A., & Yang, Y. (2017). Decellularization strategies for regenerative medicine: From processing techniques to applications. *BioMed Research International, 2017*, 9831534.

Glowacki, J., & Mizuno, S. (2008). Collagen scaffolds for tissue engineering. *Biopolymers, 89*(5), 338–344. https://doi.org/10.1002/bip.20871

Gonzalez-Pujana, A., Vining, K. H., Zhang, D. K., Santos-Vizcaino, E., Igartua, M., Hernandez, R. M., & Mooney, D. J. (2020). Multifunctional biomimetic hydrogel systems to boost the immunomodulatory potential of mesenchymal stromal cells. *Biomaterials, 257*, 120266.

Goodarzi, H., Jadidi, K., Pourmotabed, S., Sharifi, E., & Aghamollaei, H. (2019). Preparation and in vitro characterization of cross-linked collagen–gelatin hydrogel using EDC/NHS for corneal tissue engineering applications. *International Journal of Biological Macromolecules, 126*, 620–632.

Gorin, C., Rochefort, G. Y., Bascetin, R., Ying, H., Lesieur, J., Sadoine, J., et al. (2016). Priming dental pulp stem cells with fibroblast growth factor-2 increases angiogenesis of implanted tissue-engineered constructs through hepatocyte growth factor and vascular endothelial growth factor secretion. *Stem Cells Translational Medicine, 5*(3), 392–404.

Gu, S. Y., Ren, J., & Vancso, G. J. (2005). Process optimization and empirical modeling for electrospun polyacrylonitrile (PAN) nanofiber precursor of carbon nanofibers. *European Polymer Journal, 41*(11), 2559–2568.

Gu, L., Shan, T., Ma, Y., Tay, F. R., & Niu, L. (2019). Novel biomedical applications of crosslinked collagen. *Trends in Biotechnology, 37*(5), 464–491.

Hakimi, O., Mouthuy, P., & Carr, A. (2013). Synthetic and degradable patches: An emerging solution for rotator cuff repair. *International Journal of Experimental Pathology, 94*(4), 287–292.

Harvey, A. K., Thompson, M. S., Cochlin, L. E., Raju, P. A., Cui, Z., Cornell, H. R., et al. (2009). Functional imaging of tendon. *Annals British Machine Vision Association, 2009*, 1–11.

Haugh, M. G., Jaasma, M. J., & O'Brien, F. J. (2006). *Dehydrothermal crosslinking of collagen-GAG scaffolds*. Paper presented at the bioengineering in Ireland conference, 33.

Haugh, M. G., Jaasma, M. J., & O'Brien, F. J. (2009). The effect of dehydrothermal treatment on the mechanical and structural properties of collagen-GAG scaffolds. *Journal of Biomedical Materials Research Part A: An Official Journal of the Society for Biomaterials, the Japanese Society for Biomaterials, and the Australian Society for Biomaterials and the Korean Society for Biomaterials, 89*(2), 363 369.

Haugh, M. G., Murphy, C. M., McKiernan, R. C., Altenbuchner, C., & O'Brien, F. J. (2011). Crosslinking and mechanical properties significantly influence cell attachment, proliferation, and migration within collagen glycosaminoglycan scaffolds. *Tissue Engineering Part A, 17*(9-10), 1201–1208.

Hauser, R. A., & Dolan, E. E. (2011). Ligament injury and healing: An overview of current clinical concepts. *Journal of Prolotherapy, 3*(4), 836–846.

Haynl, C., Hofmann, E., Pawar, K., Förster, S., & Scheibel, T. (2016). Microfluidics-produced collagen fibers show extraordinary mechanical properties. *Nano Letters, 16*(9), 5917–5922.

He, T. (2016). *Tissue engineering 3D textiles scaffolds for multiple junction tissue regeneration*. North Carolina State University.

Heinemann, C., Brünler, R., Kreschel, C., Kruppke, B., Bernhardt, R., Aibibu, D., et al. (2019). Bioinspired calcium phosphate mineralization on net-shape-nonwoven chitosan scaffolds stimulates human bone marrow stromal cell differentiation. *Biomedical Materials (Bristol), 14*(4), 045017.

Henkel, J., Woodruff, M. A., Epari, D. R., Steck, R., Glatt, V., Dickinson, I. C., et al. (2013). Bone regeneration based on tissue engineering conceptions — A 21st century perspective. *Bone Research, 1*(3), 216–248. https://doi.org/10.4248/BR201303002

Heydarkhan-Hagvall, S., Schenke-Layland, K., Dhanasopon, A. P., Rofail, F., Smith, H., Wu, B. M., et al. (2008). Three-dimensional electrospun ECM-based hybrid scaffolds for cardiovascular tissue engineering. *Biomaterials, 29*(19), 2907–2914.

Hibino, N., Mejias, D., Pietris, N., Dean, E., Yi, T., Best, C., et al. (2015). The innate immune system contributes to tissue-engineered vascular graft performance. *The FASEB Journal, 29*(6), 2431–2438.

Homayoni, H., Ravandi, S. A. H., & Valizadeh, M. (2009). Electrospinning of chitosan nanofibers: Processing optimization. *Carbohydrate Polymers, 77*(3), 656–661.

Hong, H., Huh, M., Park, S. M., Lee, K., Kim, H. K., & Kim, D. S. (2018). Decellularized corneal lenticule embedded compressed collagen: Toward a suturable collagenous construct for limbal reconstruction. *Biofabrication, 10*(4), 045001.

Hoogenkamp, H. R., Bakker, G., Wolf, L., Suurs, P., Dunnewind, B., Barbut, S., et al. (2015). Directing collagen fibers using counter-rotating cone extrusion. *Acta Biomaterialia, 12*, 113–121.

Hu, Y., Liu, L., Dan, W., Dan, N., Gu, Z., & Yu, X. (2013). Synergistic effect of carbodiimide and dehydrothermal crosslinking on acellular dermal matrix. *International Journal of Biological Macromolecules, 55*, 221–230.

Huang, Z. M., Zhang, Y. Z., Kotaki, M., & Ramakrishna, S. (2003). A review on polymer nanofibers by electrospinning and their applications in nanocomposites. *Composites Science and Technology, 63*(15), 2223–2253.

Huang, C., Chen, S., Lai, C., Reneker, D. H., Qiu, H., Ye, Y., & Hou, H. (2006). Electrospun polymer nanofibres with small diameters. *Nanotechnology, 17*(6), 1558.

Huang, N. F., Okogbaa, J., Lee, J. C., Jha, A., Zaitseva, T. S., Paukshto, M. V., et al. (2013). The modulation of endothelial cell morphology, function, and survival using anisotropic nanofibrillar collagen scaffolds. *Biomaterials, 34*(16), 4038–4047.

Islam, A., Bohl, M. S., Tsai, A. G., Younesi, M., Gillespie, R., & Akkus, O. (2015). Biomechanical evaluation of a novel suturing scheme for grafting load-bearing collagen scaffolds for rotator cuff repair. *Clinical Biomechanics, 30*(7), 669–675.

Joanne, P., Kitsara, M., Boitard, S., Naemetalla, H., Vanneaux, V., Pernot, M., et al. (2016). Nanofibrous clinical-grade collagen scaffolds seeded with human cardiomyocytes induces cardiac remodeling in dilated cardiomyopathy. *Biomaterials, 80*, 157–168.

Johnson, G. A., Tramaglini, D. M., Levine, R. E., Ohno, K., Choi, N. Y., & Woo, S. L. (1994). Tensile and viscoelastic properties of human patellar tendon. *Journal of Orthopaedic Research, 12*(6), 796–803.

Karami, A., Tebyanian, H., Sayyad Soufdoost, R., Motavallian, E., Barkhordari, A., & Nourani, M. R. (2019). Extraction and characterization of collagen with cost-effective method from human placenta for biomedical applications. *World Journal of Plastic Surgery, 8*(3), 352–358. https://doi.org/10.29252/wjps.8.3.352

Kasimir, M., Rieder, E., Seebacher, G., Silberhumer, G., Wolner, E., Weigel, G., & Simon, P. (2003). Comparison of different decellularization procedures of porcine heart valves. *The International Journal of Artificial Organs, 26*(5), 421–427.

Kastelic, J., Galeski, A., & Baer, E. (1978). The multicomposite structure of tendon. *Connective Tissue Research, 6*(1), 11–12.

Kew, S. J., Gwynne, J. H., Enea, D., Abu-Rub, M., Pandit, A., Zeugolis, D., et al. (2011). Regeneration and repair of tendon and ligament tissue using collagen fibre biomaterials. *Acta Biomaterialia, 7*(9), 3237–3247.

Ki, C. S., Park, S. Y., Kim, H. J., Jung, H. M., Woo, K. M., Lee, J. W., & Park, Y. H. (2008). Development of 3-D nanofibrous fibroin scaffold with high porosity by electrospinning: Implications for bone regeneration. *Biotechnology Letters, 30*, 405–410.

Kim, K. W., Lee, K. H., Khil, M. S., Ho, Y. S., & Kim, H. Y. (2004). The effect of molecular weight and the linear velocity of drum surface on the properties of electrospun poly (ethylene terephthalate) nonwovens. *Fibers and Polymers, 5*, 122–127.

Kim, G. T., Lee, J. S., Shin, J. H., Ahn, Y. C., Hwang, Y. J., Shin, H. S., et al. (2005). Investigation of pore formation for polystyrene electrospun fiber: Effect of relative humidity. *Korean Journal of Chemical Engineering, 22*, 783–788.

Kim, H., Jang, J., Park, J., Lee, K. P., Lee, S., Lee, D. M., et al. (2019). Shear-induced alignment of collagen fibrils using 3D cell printing for corneal stroma tissue engineering. *Biofabrication, 11*(3), 035017.

Kin, S., Hagiwara, A., Nakase, Y., Kuriu, Y., Nakashima, S., Yoshikawa, T., et al. (2007). Regeneration of skeletal muscle using in situ tissue engineering on an acellular collagen sponge scaffold in a rabbit model. *ASAIO journal, 53*(4), 506–513.

King, M. W., Chen, J., Deshpande, M., He, T., Ramakrishna, H., Xie, Y., et al. (2019). Structural design, fabrication and evaluation of resorbable fiber-based tissue engineering scaffolds. In *Biotechnology and Bioengineering* (pp. 61–188).

Kishore, V., Uquillas, J. A., Dubikovsky, A., Alshehabat, M. A., Snyder, P. W., Breur, G. J., & Akkus, O. (2012a). In vivo response to electrochemically aligned collagen bioscaffolds. *Journal of Biomedical Materials Research Part B: Applied Biomaterials, 100*(2), 400–408.

Kishore, V., Bullock, W., Sun, X., Van Dyke, W. S., & Akkus, O. (2012b). Tenogenic differentiation of human MSCs induced by the topography of electrochemically aligned collagen threads. *Biomaterials, 33*(7), 2137–2144.

Kishore, V., Iyer, R., Frandsen, A., & Nguyen, T. (2016). In vitro characterization of electrochemically compacted collagen matrices for corneal applications. *Biomedical Materials, 11*(5), 055008.

Kitsara, M., Joanne, P., Boitard, S. E., Dhiab, I. B., Poinard, B., Menasché, P., et al. (2015). Fabrication of cardiac patch by using electrospun collagen fibers. *Microelectronic Engineering, 144*, 46–50.

Klotz, S., Danser, A. J., & Burkhoff, D. (2008). Impact of left ventricular assist device (LVAD) support on the cardiac reverse remodeling process. *Progress in Biophysics and Molecular Biology, 97*(2-3), 479–496.

Koeck, K. S., Salehi, S., Humenik, M., & Scheibel, T. (2022). Processing of continuous non-crosslinked collagen fibers for microtissue formation at the muscle-tendon Interface. *Advanced Functional Materials, 32*(15), 2112238.

Koike, N., Fukumura, D., Gralla, O., Au, P., Schechner, J. S., & Jain, R. K. (2004). Creation of long-lasting blood vessels. *Nature, 428*(6979), 138–139.

Kong, B., Chen, Y., Liu, R., Liu, X., Liu, C., Shao, Z., et al. (2020). Fiber reinforced GelMA hydrogel to induce the regeneration of corneal stroma. *Nature Communications, 11*(1), 1–12.

Koski, A., Yim, K., & Shivkumar, S. J. M. L. (2004). Effect of molecular weight on fibrous PVA produced by electrospinning. *Materials Letters, 58*(3-4), 493–497.

Koyama, H., Raines, E. W., Bornfeldt, K. E., Roberts, J. M., & Ross, R. (1996). Fibrillar collagen inhibits arterial smooth muscle proliferation through regulation of Cdk2 inhibitors. *Cell, 87*(6), 1069–1078.

Kwon, H., Brown, W. E., Lee, C. A., Wang, D., Paschos, N., Hu, J. C., & Athanasiou, K. A. (2019). Surgical and tissue engineering strategies for articular cartilage and meniscus repair. *Nature Reviews Rheumatology, 15*(9), 550–570.

Lai, E. S., Huang, N. F., Cooke, J. P., & Fuller, G. G. (2012). Aligned nanofibrillar collagen regulates endothelial organization and migration. *Regenerative Medicine, 7*(5), 649–661.

Law, J. X., Liau, L. L., Saim, A., Yang, Y., & Idrus, R. (2017). Electrospun collagen nanofibers and their applications in skin tissue engineering. *Tissue Engineering and Regenerative Medicine, 14*, 699–718.

Learn, G. D., McClellan, P. E., Knapik, D. M., Cumsky, J. L., Webster-Wood, V., Anderson, J. M., et al. (2019). Woven collagen biotextiles enable mechanically functional rotator cuff tendon regeneration during repair of segmental tendon defects in vivo. *Journal of Biomedical Materials Research Part B: Applied Biomaterials, 107*(6), 1864–1876.

Lee, J., Manoharan, V., Cheung, L., Lee, S., Cha, B., Newman, P., et al. (2019). Nanoparticle-based hybrid scaffolds for deciphering the role of multimodal cues in cardiac tissue engineering. *ACS Nano, 13*(11), 12525–12539.

León-López, A., Morales-Peñaloza, A., Martínez-Juárez, V. M., Vargas-Torres, A., Zeugolis, D. I., & Aguirre-Álvarez, G. (2019). Hydrolyzed collagen—sources and applications. *Molecules, 24*(22), 4031.

Liao, I., Moutos, F. T., Estes, B. T., Zhao, X., & Guilak, F. (2013). Composite three-dimensional woven scaffolds with interpenetrating network hydrogels to create functional synthetic articular cartilage. *Advanced Functional Materials, 23*(47), 5833–5839.

Lin, S., & Gu, L. (2015). Influence of crosslink density and stiffness on mechanical properties of type I collagen gel. *Materials, 8*(2), 551–560.

Lin, T., Wang, H., Wang, H., & Wang, X. (2004). The charge effect of cationic surfactants on the elimination of fibre beads in the electrospinning of polystyrene. *Nanotechnology, 15*(9), 1375.

Line, S., Rhodes, C., & Yamada, Y. (1993). 16 – Molecular biology of cartilage matrix. In M. Noda (Ed.), *Cellular and molecular biology of bone* (pp. 539–555). Academic. Retrieved from http://www.sciencedirect.com/science/article/pii/B9780080925004500206

Liu, H. (2019). *Collagen coated and plasma treated hernia mesh.*

Liu, R. M., Liu, Y., Forman, H. J., Olman, M., & Tarpey, M. M. (2004). Glutathione regulates transforming growth factor-β-stimulated collagen production in fibroblasts. *American Journal of Physiology-Lung Cellular and Molecular Physiology, 286*(1), L121–L128.

Liu, Y., Ramanath, H. S., & Wang, D. (2008). Tendon tissue engineering using scaffold enhancing strategies. *Trends in Biotechnology, 26*(4), 201–209.

Liu, F., Guo, R., Shen, M., Wang, S., & Shi, X. (2009). Effect of processing variables on the morphology of electrospun poly [(lactic acid)-co-(glycolic acid)] nanofibers. *Macromolecular Materials and Engineering, 294*(10), 666–672.

Liu, Y., Dong, L., Fan, J., Wang, R., & Yu, J. Y. (2011). Effect of applied voltage on diameter and morphology of ultrafine fibers in bubble electrospinning. *Journal of Applied Polymer Science, 120*(1), 592–598.

Liu, Y., Ren, L., Qi, X., Yang, L., Li, J., Wang, Y., & Zhong, J. (2014). Hydrothermal exfoliated molybdenum disulfide nanosheets as anode material for lithium ion batteries. *Journal of Energy Chemistry, 23*(2), 207–212.

Liu, Y., Xu, Y., Wang, Z., Wen, D., Zhang, W., Schmull, S., et al. (2016). Electrospun nanofibrous sheets of collagen/elastin/polycaprolactone improve cardiac repair after myocardial infarction. *American Journal of Translational Research, 8*(4), 1678.

Liu, J., Saul, D., Böker, K. O., Ernst, J., Lehman, W., & Schilling, A. F. (2018). Current methods for skeletal muscle tissue repair and regeneration. *BioMed Research International, 2018*, 1984879-11. https://doi.org/10.1155/2018/1984879

Lodish, H., Berk, A., Zipursky, S. L., Matsudaira, P., Baltimore, D., & Darnell, J. (2000). Collagen: the fibrous proteins of the matrix. *Molecular Cell Biology, 4*, 145–154.

Long, K., Liu, Y., Li, W., Wang, L., Liu, S., Wang, Y., et al. (2015). Improving the mechanical properties of collagen-based membranes using silk fibroin for corneal tissue engineering. *Journal of Biomedical Materials Research Part A, 103*(3), 1159–1168.

Lovejoy, B., Hassell, A. M., Luther, M. A., Weigl, D., & Jordan, S. R. (1994). Crystal structures of recombinant 19-kDa human fibroblast collagenase complexed to itself. *Biochemistry, 33*(27), 8207–8217.

Lu, S., Zhang, P., Sun, X., Gong, F., Yang, S., Shen, L., et al. (2013). Synthetic ePTFE grafts coated with an anti-CD133 antibody-functionalized heparin/collagen multilayer with rapid in vivo endothelialization properties. *ACS Applied Materials & Interfaces, 5*(15), 7360–7369.

Ma, J., Holden, K., Zhu, J., Pan, H., & Li, Y. (2011). The application of three-dimensional collagen-scaffolds seeded with myoblasts to repair skeletal muscle defects. *BioMed Research International, 2011*(1), 812135.

Maeda, M., Kadota, K., Kajihara, M., Sano, A., & Fujioka, K. (2001). Sustained release of human growth hormone (hGH) from collagen film and evaluation of effect on wound healing in db/db mice. *Journal of Controlled Release, 77*(3), 261–272.

Maghdouri-White, Y., Petrova, S., Sori, N., Polk, S., Wriggers, H., Ogle, R., et al. (2018). Electrospun silk–collagen scaffolds and BMP-13 for ligament and tendon repair and regeneration. *Biomedical Physics & Engineering Express, 4*(2), 025013.

Mahara, A., Somekawa, S., Kobayashi, N., Hirano, Y., Kimura, Y., Fujisato, T., & Yamaoka, T. (2015). Tissue-engineered acellular small diameter long-bypass grafts with neointima-inducing activity. *Biomaterials, 58*, 54–62.

Mammoto, A., Connor, K. M., Mammoto, T., Yung, C. W., Huh, D., Aderman, C. M., et al. (2009). A mechanosensitive transcriptional mechanism that controls angiogenesis. *Nature, 457*(7233), 1103–1108.

Manon-Jensen, T., Kjeld, N. G., & Karsdal, M. A. (2016). Collagen-mediated hemostasis. *Journal of Thrombosis and Haemostasis, 14*(3), 438–448.

Matinong, A. M. E., Chisti, Y., Pickering, K. L., & Haverkamp, R. G. (2022). Collagen extraction from animal skin. *Biology, 11*(6), 905.

Matthews, J. A., Wnek, G. E., Simpson, D. G., & Bowlin, G. L. (2002). Electrospinning of collagen nanofibers. *Biomacromolecules, 3*(2), 232–238.

Matthews, J. A., Boland, E. D., Wnek, G. E., Simpson, D. G., & Bowlin, G. L. (2016). Electrospinning of collagen type II: A feasibility study. *Journal of Bioactive and Compatible Polymers*. https://doi.org/10.1177/0883911503018002003

McAllister, T. N., Maruszewski, M., Garrido, S. A., Wystrychowski, W., Dusserre, N., Marini, A., et al. (2009). Effectiveness of haemodialysis access with an autologous tissue-engineered vascular graft: a multicentre cohort study. *The Lancet, 373*(9673), 1440–1446.

McCullen, S. D., Haslauer, C. M., & Loboa, E. G. (2010). Fiber-reinforced scaffolds for tissue engineering and regenerative medicine: Use of traditional textile substrates to nanofibrous arrays. *Journal of Materials Chemistry, 20*(40), 8776–8788. https://doi.org/10.1039/c0jm01443e

Medeiros, E. S., Mattoso, L. H., Offeman, R. D., Wood, D. F., & Orts, W. J. (2008). Effect of relative humidity on the morphology of electrospun polymer fibers. *Canadian Journal of Chemistry, 86*(6), 590–599.

Megelski, S., Stephens, J. S., Chase, D. B., & Rabolt, J. F. (2002). Micro- and nanostructured surface morphology on electrospun polymer fibers. *Macromolecules, 35*(22), 8456–8466. https://doi.org/10.1021/ma020444a

Mekhail, M., Wong, K. K. H., Padavan, D. T., Wu, Y., O'Gorman, D. B., & Wan, W. (2011). Genipin-cross-linked electrospun collagen fibers. *Journal of Biomaterials Science, Polymer Edition, 22*(17), 2241–2259.

Menon, S. N., Flegg, J. A., McCue, S. W., Schugart, R. C., Dawson, R. A., & McElwain, D. S. (2012). Modelling the interaction of keratinocytes and fibroblasts during normal and abnormal wound healing processes. *Proceedings of the Royal Society B: Biological Sciences, 279*(1741), 3329–3338.

Meyer, M. (2019). Processing of collagen based biomaterials and the resulting materials properties. *Biomedical Engineering Online, 18*(1), 24. https://doi.org/10.1186/s12938-019-0647-0

Mit-uppatham, C., Nithitanakul, M., & Supaphol, P. (2004). Ultrafine electrospun polyamide-6 fibers: effect of solution conditions on morphology and average fiber diameter. *Macromolecular Chemistry and Physics, 205*(17), 2327–2338.

Mo, X. M., Xu, C. Y., Kotaki, M. E. A., & Ramakrishna, S. (2004). Electrospun P (LLA-CL) nanofiber: a biomimetic extracellular matrix for smooth muscle cell and endothelial cell proliferation. *Biomaterials, 25*(10), 1883–1890.

Montero, P., Flandes-Iparraguirre, M., Musquiz, S., Pérez Araluce, M., Plano, D., Sanmartín, C., ... & Mazo, M. M. (2020). Cells, materials, and fabrication processes for cardiac tissue engineering. *Frontiers in Bioengineering and Biotechnology, 8*, 955

Mostafizur Rahman, M., Shamsuzzaman, M., Das, D., Abdus Shahid, M., & Hoque, M. B. (2023). Introduction to textiles and textile fibers. In *Advanced technology in textiles: Fibre to apparel* (pp. 1–29). Springer.

Motta, F. N., Azevedo, C., Araujo, C. N., Santana, J. S., & Bastos, I. M. (2012). *Degradation of tipe I collagen and pathogenesis of infectious diseases*. Laboratory of Pathogen-Host Interface, Department of Cell Biology, the University of Brasilia.

Moura, L. I., Dias, A. M., Suesca, E., Casadiegos, S., Leal, E. C., Fontanilla, M. R., et al. (2014). Neurotensin-loaded collagen dressings reduce inflammation and improve wound healing in diabetic mice. *Biochimica et Biophysica Acta (BBA) - Molecular Basis of Disease, 1842*(1), 32–43.

Nakayama, K. H., Hong, G., Lee, J. C., Patel, J., Edwards, B., Zaitseva, T. S., et al. (2015). Aligned-braided nanofibrillar scaffold with endothelial cells enhances arteriogenesis. *ACS Nano, 9*(7), 6900–6908.

Nakayama, K. H., Quarta, M., Paine, P., Alcazar, C., Karakikes, I., Garcia, V., et al. (2019). Treatment of volumetric muscle loss in mice using nanofibrillar scaffolds enhances vascular organization and integration. *Communications Biology, 2*(1), 1–16.

Nezarati, R. M., Eifert, M. B., & Cosgriff-Hernandez, E. (2013). Effects of humidity and solution viscosity on electrospun fiber morphology. *Tissue Engineering Part C: Methods, 19*(10), 810–819.

Nielsen, M. J., & Karsdal, M. A. (2016). Chapter 3 – Type III collagen. In M. A. Karsdal (Ed.), *Biochemistry of collagens, laminins and elastin* (pp. 21–30). Academic.

Niklason, L. E., & Lawson, J. H. (2020). Bioengineered human blood vessels. *Science, 370*(6513), eaaw8682.

Pabbruwe, M. B., Kafienah, W., Tarlton, J. F., Mistry, S., Fox, D. J., & Hollander, A. P. (2010). Repair of meniscal cartilage white zone tears using a stem cell/collagen-scaffold implant. *Biomaterials, 31*(9), 2583–2591.

Parenteau-Bareil, R., Gauvin, R., & Berthod, F. (2010). Collagen-based biomaterials for tissue engineering applications. *Materials, 3*(3), 1863–1887.

Park, H., Radisic, M., Lim, J. O., Chang, B. H., & Vunjak-Novakovic, G. (2005). A novel composite scaffold for cardiac tissue engineering. *In Vitro Cellular & Developmental Biology-Animal, 41*(7), 188–196.

Park, K. H., Kwon, J. B., Park, J. H., Shin, J. C., Han, S. H., & Lee, J. W. (2019). Collagen dressing in the treatment of diabetic foot ulcer: A prospective, randomized, placebo-controlled, single-center study. *Diabetes Research and Clinical Practice, 156*, 107861.

Pham, Q. P., Sharma, U., & Mikos, A. G. (2006). Electrospinning of polymeric nanofibers for tissue engineering applications: A review. *Tissue Engineering, 12*(5), 1197–1211.

Pollot, B. E., Rathbone, C. R., Wenke, J. C., & Guda, T. (2018). Natural polymeric hydrogel evaluation for skeletal muscle tissue engineering. *Journal of Biomedical Materials Research. Part B, Applied Biomaterials, 106*(2), 672–679. https://doi.org/10.1002/jbm.b.33859

Prabhakaran, M. P., Kai, D., Ghasemi-Mobarakeh, L., & Ramakrishna, S. (2011). Electrospun biocomposite nanofibrous patch for cardiac tissue engineering. *Biomedical Materials, 6*(5), 055001.

Puetzer, J. L., Ma, T., Sallent, I., Gelmi, A., & Stevens, M. M. (2021). Driving hierarchical collagen fiber formation for functional tendon, ligament, and meniscus replacement. *Biomaterials, 269*, 120527.

Qian, S., Wang, Z., Zheng, Z., Ran, J., Zhu, J., & Chen, W. (2019). A collagen and silk scaffold for improved healing of the tendon and bone interface in a rabbit model. Medical Science Monitor: International Medical Journal of Experimental and Clinical Research, 25, 269.

Qiao, Z., Lian, M., Han, Y., Sun, B., Zhang, X., Jiang, W., et al. (2021). Bioinspired stratified electrowritten fiber-reinforced hydrogel constructs with layer-specific induction capacity for functional osteochondral regeneration. *Biomaterials, 266*, 120385.

Quarta, M., Cromie, M., Chacon, R., Blonigan, J., Garcia, V., Akimenko, I., et al. (2017). Bioengineered constructs combined with exercise enhance stem cell-mediated treatment of volumetric muscle loss. *Nature Communications, 8*(1), 15613.

Rafat, M., Li, F., Fagerholm, P., Lagali, N. S., Watsky, M. A., Munger, R., et al. (2008). PEG-stabilized carbodiimide crosslinked collagen–chitosan hydrogels for corneal tissue engineering. *Biomaterials, 29*(29), 3960–3972.

Ramakrishna, H. (2019). *Biodegradable biphasic scaffolds using braiding technology and surface treatments for regeneration of tendon-bone junction tissue*. North Carolina State University.

Rangaraj, A., Harding, K., & Leaper, D. (2011). Role of collagen in wound management. *Wounds UK, 7*(2), 54–63.

Ratcliffe, A., Butler, D. L., Dyment, N. A., Cagle, P. J., Proctor, C. S., Ratcliffe, S. S., & Flatow, E. L. (2015). Scaffolds for tendon and ligament repair and regeneration. *Annals of Biomedical Engineering, 43*(3), 819–831.

Rath, G., Hussain, T., Chauhan, G., Garg, T., & Goyal, A. K. (2016). Collagen nanofiber containing silver nanoparticles for improved wound-healing applications. *Journal of Drug Targeting, 24*(6), 520–529.

Relaix, F., & Zammit, P. S. (2012). Satellite cells are essential for skeletal muscle regeneration: The cell on the edge returns centre stage. *Development (Cambridge), 139*(16), 2845–2856. https://doi.org/10.1242/dev.069088

Rho, K. S., Jeong, L., Lee, G., Seo, B., Park, Y. J., Hong, S., et al. (2006). Electrospinning of collagen nanofibers: Effects on the behavior of normal human keratinocytes and early-stage wound healing. *Biomaterials, 27*(8), 1452–1461.

Rieu, C., Picaut, L., Mosser, G., & Trichet, L. (2017). From tendon injury to collagen-based tendon regeneration: Overview and recent advances. *Current Pharmaceutical Design, 23*(24), 3483–3506.

Rimoin, D. L., Lachman, R., & Unger, S. (2013). Chapter 158 – Chondrodysplasias. In D. Rimoin, R. Pyeritz, & B. Korf (Eds.), *Emery and rimoin's principles and practice of medical genetics* (pp. 1–45). Academic.

Rittié, L. (2017). Type I collagen purification from rat tail tendons. In *Fibrosis* (pp. 287–308). Springer.

Rodoplu, D., & Mutlu, M. (2012). Effects of electrospinning setup and process parameters on nanofiber morphology intended for the modification of quartz crystal microbalance surfaces. *Journal of Engineered Fibers and Fabrics, 7*(2), 155892501200700217.

Roeder, R. A., Lantz, G. C., & Geddes, L. A. (2001). Mechanical remodeling of small-intestine submucosa small-diameter vascular grafts—A preliminary report. *Biomedical Instrumentation & Technology, 35*(2), 110–120.

Rogers, J. G., Bostic, R. R., Tong, K. B., Adamson, R., Russo, M., & Slaughter, M. S. (2012). Cost-effectiveness analysis of continuous-flow left ventricular assist devices as destination therapy. *Circulation: Heart Failure, 5*(1), 10–16.

Rosenthal, F. M., & Köhler, G. (1997). Collagen as matrix for neo-organ formation by gene-transfected fibroblasts. *Anticancer Research, 17*(2A), 1179–1186.

Rouwkema, J., & Khademhosseini, A. (2016). Vascularization and angiogenesis in tissue engineering: Beyond creating static networks. *Trends in Biotechnology, 34*(9), 733–745.

Sand, J. M. B., Genovese, F., Gudmann, N. S., & Karsdal, M. A. (2019). Chapter 4 – Type IV collagen. In M. A. Karsdal (Ed.), *Biochemistry of collagens, laminins and elastin* (2nd ed., pp. 37–49). Academic.

Sandusky, G. E., Lantz, G. C., & Badylak, S. F. (1995). Healing comparison of small intestine submucosa and ePTFE grafts in the canine carotid artery. *Journal of Surgical Research, 58*(4), 415–420.

Sano, H., Hatta, T., Yamamoto, N., & Itoi, E. (2013). Stress distribution within rotator cuff tendons with a crescent-shaped and an L-shaped tear. *The American Journal of Sports Medicine, 41*(10), 2262–2269.

Schipani, R., Scheurer, S., Florentin, R., Critchley, S. E., & Kelly, D. J. (2020). Reinforcing interpenetrating network hydrogels with 3D printed polymer networks to engineer cartilage mimetic composites. *Biofabrication, 12*(3), 035011.

Schutte, S. C., Chen, Z., Brockbank, K. G., & Nerem, R. M. (2010). Cyclic strain improves strength and function of a collagen-based tissue-engineered vascular media. *Tissue Engineering Part A, 16*(10), 3149–3157.

Seifu, D. G., Purnama, A., Mequanint, K., & Mantovani, D. (2013). Small-diameter vascular tissue engineering. *Nature Reviews Cardiology, 10*(7), 410–421.

Şener, A. G., Altay, A. S., & Altay, F. (2011, December). Effect of voltage on morphology of electrospun nanofibers. In *2011 7th international conference on Electrical and Electronics Engineering (ELECO)* (pp. 1–324). IEEE.

Sensini, A., & Cristofolini, L. (2018). Biofabrication of electrospun scaffolds for the regeneration of tendons and ligaments. *Materials, 11*(10), 1963.

Serpooshan, V., Zhao, M., Metzler, S. A., Wei, K., Shah, P. B., Wang, A., et al. (2013). The effect of bioengineered acellular collagen patch on cardiac remodeling and ventricular function post myocardial infarction. *Biomaterials, 34*(36), 9048–9055.

Shah, R., Knowles, J. C., Hunt, N. P., & Lewis, M. P. (2016). Development of a novel smart scaffold for human skeletal muscle regeneration. *Journal of Tissue Engineering and Regenerative Medicine, 10*(2), 162–171.

Shen, Y., Xu, Y., Yi, B., Wang, X., Tang, H., Chen, C., & Zhang, Y. (2021). Engineering a highly biomimetic chitosan-based cartilage scaffold by using short fibers and a cartilage-decellularized matrix. *Biomacromolecules, 22*(5), 2284–2297.

Shimomura, A., Shiino, S., Kawauchi, J., Takizawa, S., Sakamoto, H., Matsuzaki, J., et al. (2016). Novel combination of serum microRNA for detecting breast cancer in the early stage. *Cancer Science, 107*(3), 326–334.

Shojaee, M., Wood, K. B., Moore, L. K., & Bashur, C. A. (2017). Peritoneal pre-conditioning reduces macrophage marker expression in collagen-containing engineered vascular grafts. *Acta Biomaterialia, 64*, 80–93.

Shojaee, M., Swaminathan, G., Bashur, C. A., & Ramamurthi, A. (2018). Temporal changes in peritoneal cell phenotype and neoelastic matrix induction with hyaluronan oligomers and TGF-β1 after implantation of engineered conduits. *Journal of Tissue Engineering and Regenerative Medicine, 12*(6), 1420–1431.

Shoulders, M. D., & Raines, R. T. (2009). Collagen structure and stability. *Annual Review of Biochemistry, 78*, 929–958.

Silvipriya, K. S., Kumar, K. K., Bhat, A. R., Kumar, B. D., John, A., & Lakshmanan, P. (2015). Collagen: Animal sources and biomedical application. *Journal of Applied Pharmaceutical Science, 5*(3), 123–127.

Sirivisoot, S., Pareta, R., & Harrison, B. S. (2014). Protocol and cell responses in three-dimensional conductive collagen gel scaffolds with conductive polymer nanofibres for tissue regeneration. *Interface Focus, 4*(1), 20130050. https://doi.org/10.1098/rsfs.2013.0050

Siriwardane, M. L., DeRosa, K., Collins, G., & Pfister, B. J. (2014). Controlled formation of cross-linked collagen fibers for neural tissue engineering applications. *Biofabrication, 6*(1), 015012.

Sogorkova, J., Zapotocky, V., Cepa, M., Stepankova, V., Vagnerova, H., Batova, J., et al. (2018). Optimization of cell growth on palmitoyl-hyaluronan knitted scaffolds developed for tissue engineering applications. *Journal of Biomedical Materials Research. Part A, 106*(6), 1488–1499. https://doi.org/10.1002/jbm.a.36353

Solouk, A., Cousins, B. G., Mirzadeh, H., & Seifalian, A. M. (2011). Application of plasma surface modification techniques to improve hemocompatibility of vascular grafts: A review. *Biotechnology and Applied Biochemistry, 58*(5), 311–327. https://doi.org/10.1002/bab.50

Somers, S., Spector, A., DiGirolamo, D., & Grayson, W. L. (2017). Biophysical stimulation for engineering functional skeletal muscle. *Tissue Engineering. Part B, Reviews, 23*(ja), 362–372. https://doi.org/10.1089/ten.TEB.2016.0444

Strawich, E., & Nimni, M. E. (1971). Properties of a collagen molecule containing three identical components extracted from bovine articular cartilage. *Biochemistry, 10*(21), 3905–3911.

Sugiura, T., Agarwal, R., Tara, S., Yi, T., Lee, Y. U., Breuer, C. K., ... & Shinoka, T. (2017). Tropoelastin inhibits intimal hyperplasia of mouse bioresorbable arterial vascular grafts. *Acta Biomaterialia, 52*, 74–80.

Sukigara, S., Gandhi, M., Ayutsede, J., Micklus, M., & Ko, F. (2003). Regeneration of Bombyx mori silk by electrospinning—part 1: Processing parameters and geometric properties. *Polymer, 44*(19), 5721–5727.

Sun, X., Altalhi, W., & Nunes, S. S. (2016). Vascularization strategies of engineered tissues and their application in cardiac regeneration. *Advanced Drug Delivery Reviews, 96*, 183–194.

Suntornnond, R., An, J., & Chua, C. K. (2016). Effect of gas plasma on polycaprolactone (PCL) membrane wettability and collagen type I immobilized for enhancing cell proliferation. *Materials Letters, 171*, 293–296.

Tallant, C., Marrero, A., & Gomis-Rüth, F. X. (2010). Matrix metalloproteinases: Fold and function of their catalytic domains. *Biochimica Et Biophysica Acta (BBA)-Molecular Cell Research, 1803*(1), 20–28.

Tao, J., & Shivkumar, S. (2007). Molecular weight dependent structural regimes during the electrospinning of PVA. *Materials Letters, 61*(11-12), 2325–2328.

Taraballi, F., Zanini, S., Lupo, C., Panseri, S., Cunha, C., Riccardi, C., et al. (2013). Amino and carboxyl plasma functionalization of collagen films for tissue engineering applications. *Journal of Colloid and Interface Science, 394*, 590–597.

Tashakori-Miyanroudi, M., Rakhshan, K., Ramez, M., Asgarian, S., Janzadeh, A., Azizi, Y., et al. (2020). Conductive carbon nanofibers incorporated into collagen bio-scaffold assists myocardial injury repair. *International Journal of Biological Macromolecules, 163*, 1136–1146.

Thon, S. G., O'Malley, L., O'Brien, M. J., & Savoie, F. H., III. (2019). Evaluation of healing rates and safety with a bioinductive collagen patch for large and massive rotator cuff tears: 2-year safety and clinical outcomes. *The American Journal of Sports Medicine, 47*(8), 1901–1908.

Thorpe, C. T., Birch, H. L., Clegg, P. D., & Screen, H. R. (2013). The role of the non-collagenous matrix in tendon function. *International Journal of Experimental Pathology, 94*(4), 248–259.

Thorpe, C. T., Birch, H. L., Clegg, P. D., & Screen, H. R. (2015). Tendon physiology and mechanical behavior: Structure–function relationships. In *Tendon regeneration* (pp. 3–39). Academic.

Tillman, B. W., Yazdani, S. K., Lee, S. J., Geary, R. L., Atala, A., & Yoo, J. J. (2009). The in vivo stability of electrospun polycaprolactone-collagen scaffolds in vascular reconstruction. *Biomaterials, 30*(4), 583–588.

Toba, T., Nakamura, T., Matsumoto, K., Fukuda, S., Yoshitani, M., Ueda, H., et al. (2002). Influence of dehydrothermal crosslinking on the growth of PC-12 cells cultured on laminin coated collagen. *Asaio Journal, 48*(1), 17–20.

Tonndorf, R., Aibibu, D., & Cherif, C. (2020). Collagen multifilament spinning. *Materials Science and Engineering: C, 106*, 110105.

Trappmann, B., Gautrot, J. E., Connelly, J. T., Strange, D. G., Li, Y., Oyen, M. L., et al. (2012). Extracellular-matrix tethering regulates stem-cell fate. *Nature Materials, 11*(7), 642–649.

Uquillas, J. A., Kishore, V., & Akkus, O. (2012). Genipin crosslinking elevates the strength of electrochemically aligned collagen to the level of tendons. *Journal of the Mechanical Behavior of Biomedical Materials, 15*, 176–189.

Van Kampen, C., Arnoczky, S., Parks, P., Hackett, E., Ruehlman, D., Turner, A., & Schlegel, T. (2013). Tissue-engineered augmentation of a rotator cuff tendon using a reconstituted collagen scaffold: A histological evaluation in sheep. *Muscles, Ligaments and Tendons Journal, 3*(3), 229.

Vasireddi, R., Kruse, J., Vakili, M., Kulkarni, S., Keller, T. F., Monteiro, D. C., & Trebbin, M. (2019). Solution blow spinning of polymer/nanocomposite micro-/nanofibers with tunable diameters and morphologies using a gas dynamic virtual nozzle. *Scientific Reports, 9*(1), 1–10.

Veleirinho, B., Rei, M. F., & Lopes-DA-Silva, J. A. (2008). Solvent and concentration effects on the properties of electrospun poly (ethylene terephthalate) nanofiber mats. *Journal of Polymer Science Part B: Polymer Physics, 46*(5), 460–471.

Verma, R. P., & Hansch, C. (2007). Matrix metalloproteinases (MMPs): Chemical–biological functions and (Q) SARs. *Bioorganic & Medicinal Chemistry, 15*(6), 2223–2268.

Visser, J., Melchels, F. P., Jeon, J. E., Van Bussel, E. M., Kimpton, L. S., Byrne, H. M., et al. (2015). Reinforcement of hydrogels using three-dimensionally printed microfibres. *Nature Communications, 6*(1), 1–10.

von der Mark, K., Park, J., Bauer, S., & Schmuki, P. (2010). Nanoscale engineering of biomimetic surfaces: Cues from the extracellular matrix. *Cell and Tissue Research, 339*(1), 131–153.

Vunjak-Novakovic, G., Tandon, N., Godier, A., Maidhof, R., Marsano, A., Martens, T. P., & Radisic, M. (2010). Challenges in cardiac tissue engineering. *Tissue Engineering Part B: Reviews, 16*(2), 169–187.

Wang, I. N. E., Mitroo, S., Chen, F. H., Lu, H. H., & Doty, S. B. (2006). Age-dependent changes in matrix composition and organization at the ligament-to-bone insertion. *Journal of Orthopaedic Research, 24*(8), 1745–1755.

Wang, C., Chien, H. S., Hsu, C. H., Wang, Y. C., Wang, C. T., & Lu, H. A. (2007). Electrospinning of polyacrylonitrile solutions at elevated temperatures. *Macromolecules, 40*(22), 7973–7983.

Wang, C., Brisson, B. K., Terajima, M., Li, Q., Han, B., Goldberg, A. M., et al. (2020). Type III collagen is a key regulator of the collagen fibrillar structure and biomechanics of articular cartilage and meniscus. *Matrix Biology, 85*, 47–67.

Wang, X., Chung, L., Hooks, J., Maestas, D. R., Lebid, A., Andorko, J. I., et al. (2021). Type 2 immunity induced by bladder extracellular matrix enhances corneal wound healing. *Science. Advances, 7*(16), eabe2635.

Weadock, K. S., Miller, E. J., Bellincampi, L. D., Zawadsky, J. P., & Dunn, M. G. (1995). Physical crosslinking of collagen fibers: comparison of ultraviolet irradiation and dehydrothermal treatment. *Journal of Biomedical Materials Research, 29*(11), 1373–1379.

Wessels, Q. B., & Pretorius, E. (2008). Enhanced stabilization of collagen-based dermal regeneration scaffolds through the combination of physical and chemical crosslinking. *South African Journal of Science, 104*(11-12), 513–516.

Wihodo, M., & Moraru, C. I. (2013). Physical and chemical methods used to enhance the structure and mechanical properties of protein films: A review. *Journal of Food Engineering, 114*(3), 292–302.

Woo, S. L., Chan, S. S., & Yamaji, T. (1997). Biomechanics of knee ligament healing, repair and reconstruction. *Journal of Biomechanics, 30*(5), 431–439.

Woo, S. L. Y., Abramowitch, S. D., Kilger, R., & Liang, R. (2006). Biomechanics of knee ligaments: Injury, healing, and repair. *Journal of Biomechanics, 39*(1), 1–20.

Xiang, Z., Liao, R., Kelly, M. S., & Spector, M. (2006). Collagen–GAG scaffolds grafted onto myocardial infarcts in a rat model: A delivery vehicle for mesenchymal stem cells. *Tissue Engineering, 12*(9), 2467–2478.

Xie, Y. (2018). *An advanced collagen yarn based textile patch for rotator cuff tendon repair*. North Carolina State University.

Xie, Y., Guan, Y., Kim, S., & King, M. W. (2016). The mechanical performance of weft-knitted/electrospun bilayer small diameter vascular prostheses. *Journal of the Mechanical Behavior of Biomedical Materials, 61*, 410–418.

Xie, Y., Chen, J., Celik, H., Akkus, O., & King, M. W. (2021). Evaluation of an electrochemically aligned collagen yarn for textile scaffold fabrication. *Biomedical Materials, 16*(2), 025001.

Xie, Y., Zhang, F., Akkus, O., & King, M. W. (2022). A collagen/PLA hybrid scaffold supports tendon-derived cell growth for tendon repair and regeneration. *Journal of Biomedical Materials Research Part B: Applied Biomaterials, 110*(12), 2624–2635.

Xu, Y., Li, L., Wang, H., Yu, X., Gu, Z., Huang, C., & Peng, H. (2013). In vitro cytocompatibility evaluation of alginate dialdehyde for biological tissue fixation. *Carbohydrate Polymers, 92*(1), 448–454.

Yang, Q., Li, Z., Hong, Y., Zhao, Y., Qiu, S., Wang, C. E., & Wei, Y. (2004). Influence of solvents on the formation of ultrathin uniform poly (vinyl pyrrolidone) nanofibers with electrospinning. *Journal of Polymer Science Part B: Polymer Physics, 42*(20), 3721–3726.

Yang, S., Shi, X., Li, X., Wang, J., Wang, Y., & Luo, Y. (2019). Oriented collagen fiber membranes formed through counter-rotating extrusion and their application in tendon regeneration. *Biomaterials, 207*, 61–75.

Yokota, T., McCourt, J., Ma, F., Ren, S., Li, S., Kim, T. H., ... & Deb, A. (2020). Type V collagen in scar tissue regulates the size of scar after heart injury. *Cell, 182*(3), 545–562.

Yoon, J. H., & Halper, J. (2005). Tendon proteoglycans: Biochemistry and function. *Journal of Musculoskeletal & Neuronal Interactions, 5*(1), 22–34.

Yoon, Y., Kim, C. H., Lee, J. E., Yoon, J., Lee, N. K., Kim, T. H., & Park, S. (2019). 3D bioprinted complex constructs reinforced by hybrid multilayers of electrospun nanofiber sheets. *Biofabrication, 11*(2), 025015.

Younesi, M., Islam, A., Kishore, V., Anderson, J. M., & Akkus, O. (2014). Tenogenic induction of human MSCs by anisotropically aligned collagen biotextiles. *Advanced Functional Materials, 24*(36), 5762–5770.

Yu, X., Tang, C., Xiong, S., Yuan, Q., Gu, Z., Li, Z., & Hu, Y. (2016). Modification of collagen for biomedical applications: A review of physical and chemical methods. *Current Organic Chemistry, 20*(17), 1797–1812.

Zeugolis, D. I., Li, B., Lareu, R. R., Chan, C. K., & Raghunath, M. (2008a). Collagen solubility testing, a quality assurance step for reproducible electro-spun nano-fibre fabrication. A technical note. *Journal of Biomaterials Science, Polymer Edition, 19*(10), 1307–1317.

Zeugolis, D. I., Khew, S. T., Yew, E. S., Ekaputra, A. K., Tong, Y. W., Yung, L. L., et al. (2008b). Electro-spinning of pure collagen nano-fibres–just an expensive way to make gelatin? *Biomaterials, 29*(15), 2293–2305.

Zhang, F., & King, M. W. (2022). Immunomodulation strategies for the successful regeneration of a tissue-engineered vascular graft. *Advanced Healthcare Materials, 11*(12), 2200045.

Zhang, S., Wang, D., Estrov, Z., Raj, S., Willerson, J. T., & Yeh, E. T. (2004). Both cell fusion and transdifferentiation account for the transformation of human peripheral blood CD34-positive cells into cardiomyocytes in vivo. *Circulation, 110*(25), 3803–3807.

Zhang, C., Yuan, X., Wu, L., Han, Y., & Sheng, J. (2005). Study on morphology of electrospun poly (vinyl alcohol) mats. *European Polymer Journal, 41*(3), 423–432.

Zhang, X., Chen, X., Yang, T., Zhang, N., Dong, L., Ma, S., et al. (2014). The effects of different crossing-linking conditions of genipin on type I collagen scaffolds: An in vitro evaluation. *Cell and Tissue Banking, 15*(4), 531–541.

Zhang, Y., Ran, L., Li, C., & Chen, X. (2015). Diversity, structures, and collagen-degrading mechanisms of bacterial collagenolytic proteases. *Applied and Environmental Microbiology, 81*(18), 6098–6107.

Zhang, J., Sisley, A. M., Anderson, A. J., Taberner, A. J., McGhee, C. N., & Patel, D. V. (2016). Characterization of a novel collagen scaffold for corneal tissue engineering. *Tissue Engineering Part C: Methods, 22*(2), 165–172.

Zhang, C., Du, L., Sun, P., Shen, L., Zhu, J., Pang, K., & Wu, X. (2017). Construction of tissue-engineered full-thickness cornea substitute using limbal epithelial cell-like and corneal endothelial cell-like cells derived from human embryonic stem cells. *Biomaterials, 124*, 180–194.

Zhang, F., Xie, Y., Celik, H., Akkus, O., Bernacki, S. H., & King, M. W. (2019). Engineering small-caliber vascular grafts from collagen filaments and nanofibers with comparable mechanical properties to native vessels. *Biofabrication, 11*(3), 035020.

Zhang, F., Bambharoliya, T., Xie, Y., Liu, L., Celik, H., Wang, L., et al. (2021). A hybrid vascular graft harnessing the superior mechanical properties of synthetic fibers and the biological performance of collagen filaments. *Materials Science and Engineering: C, 118*, 111418.

Zhang, F., Gluck, J. M., Brown, A. C., Zaharoff, D. A., & King, M. W. (2023a). Heparin affinity-based IL-4 delivery to modulate macrophage phenotype and endothelial cell activity in vitro. *ACS Applied Materials & Interfaces, 15*(23), 27457–27470.

Zhang, F., Tao, H., Gluck, J. M., Wang, L., Daneshmand, M. A., & King, M. W. (2023b). A textile-reinforced composite vascular graft that modulates macrophage polarization and enhances endothelial cell migration, adhesion and proliferation in vitro. *Soft Matter, 19*(8), 1624–1641.

Zheng, J., He, A., Li, J., Xu, J., & Han, C. C. (2006). Studies on the controlled morphology and wettability of polystyrene surfaces by electrospinning or electrospraying. *Polymer, 47*(20), 7095–7102.

Zheng, Z., Ran, J., Chen, W., Hu, Y., Zhu, T., Chen, X., et al. (2017). Alignment of collagen fiber in knitted silk scaffold for functional massive rotator cuff repair. *Acta Biomaterialia, 51*, 317–329.

Zhou, H., Green, T. B., & Joo, Y. L. (2006). The thermal effects on electrospinning of polylactic acid melts. *Polymer, 47*(21), 7497–7505.

Zhou, T., Wang, N., Xue, Y., Ding, T., Liu, X., Mo, X., & Sun, J. (2016). Electrospun tilapia collagen nanofibers accelerating wound healing via inducing keratinocytes proliferation and differentiation. *Colloids and Surfaces B: Biointerfaces, 143*, 415–422.

Zitnay, J. L., Reese, S. P., Tran, G., Farhang, N., Bowles, R. D., & Weiss, J. A. (2018). Fabrication of dense anisotropic collagen scaffolds using biaxial compression. *Acta Biomaterialia, 65*, 76–87.

Index

F. Zhang et al., *Collagen-Based Biotextiles for Tissue Engineering*,
https://doi.org/10.1007/978-3-031-71563-1

The manufacturer's authorised representative in the EU is Springer Nature Customer Service Centre GmbH, Europaplatz 3, 69115 Heidelberg, Germany. If you have any concerns regarding our products, please contact ProductSafety@springernature.com

Printed and bound by CPI Group (UK) Ltd, Croydon, CR0 4YY
01/10/2026
02225944-0004